Elements of Justice

What is justice? Questions of justice are questions about what people are due, but what that means in practice depends on context. Depending on context, the formal question of what people are due is answered by principles of desert, reciprocity, equality, or need. Justice, thus, is a constellation of elements that exhibit a degree of integration and unity, but the integrity of justice is limited, in a way that is akin to the integrity of a neighborhood. A *theory* of justice is a map of that neighborhood.

David Schmidtz is Professor of Philosophy, joint Professor of Economics, and Director of the Program in Philosophy of Freedom at the University of Arizona. He is the author of *Rational Choice and Moral Agency* and coauthor, with Robert Goodin, of *Social Welfare and Individual Responsibility*. He is editor of *Robert Nozick* and coeditor, with Elizabeth Willott, of *Environmental Ethics: What Really Matters, What Really Works*. His lectures on justice have taken him to twenty countries and six continents.

Elements of Justice

DAVID SCHMIDTZ

University of Arizona

CAMBRIDGE
UNIVERSITY PRESS

32 Avenue of the Americas, New York NY 10013-2473, USA

Cambridge University Press is part of the University of Cambridge.

It furthers the University's mission by disseminating knowledge in the pursuit of education, learning and research at the highest international levels of excellence.

www.cambridge.org
Information on this title: www.cambridge.org/9780521831642

First published 2006
Reprinted 2007

A catalogue record for this publication is available from the British Library

Library of Congress Cataloguing in Publication data

Schmidtz, David.
Elements of Justice / David Schmidtz.
p. cm.
Includes bibliographical references and index.
ISBN 0-521-83164-4 (hardback) – ISBN 0-521-53936-6 (pbk.)
1. Justice (Philosophy) I. Title.
B105.J87S35 2006
172′.2-dc21 2005011489

ISBN 978-0-521-83164-2 Hardback

Contents

PART 4 EQUAL RESPECT AND EQUAL SHARES

PART 5 MEDITATIONS ON NEED

PART 6 THE RIGHT TO DISTRIBUTE

Acknowledgments

Whenever I would run into James Rachels at a conference, he always seemed acutely aware of how much fun it is to be a philosopher. I could not match the masterful simplicity of Jim's introductory text, *The Elements of Moral Philosophy*, but I did pretty much borrow his title, conceiving the tribute before I learned he was dying of bladder cancer. To my astonishment, Jim e-mailed me from his hospital bed only days before he died, saying one of his few regrets was not getting to know me better. I have no idea how many such e-mails Jim sent, but that's the kind of man he was, thoughtful and in love with life, no matter what.

I want to thank Marty Zupan for inviting me to a fundraiser in Palm Beach in February of 2003. I thank Randy Kendrick, whom I met in Palm Beach, for calling a week later to invite Elizabeth and me to dinner with her and her husband Ken in Phoenix. I declined, saying I had been diagnosed with a brain tumor two days before and was not feeling very social. Randy demanded that I consult her friend, Dr. Robert Spetzler. As one neurosurgeon described Spetzler, it is hard to explain what makes one pianist merely excellent and the next one a virtuoso, but Spetzler is a virtuoso. His patients simply do better than other people's patients. So, I thank Dr. Spetzler. Even as brain surgeries go, this was a delicate procedure. I may well have died, or survived as a shell, if not for him.

In the aftermath, Kit Wellman and John Tomasi, among many others, called to ask whether there was anything they could do. I probably was supposed to say, "No thanks, it's enough that you called, but if I think of anything . . ." Instead, emboldened by awareness that life is indeed short, I said, "How about a workshop on my book?" I'm especially grateful to Kit, John, and Dave Estlund for putting those events together. At the Georgia

State Workshop, Andrew Altman, Andrew I. Cohen, Bill Edmundson, George Rainbolt, Geoff Sayre-McCord, and Kit Wellman served as commentators. Alex Kaufman and Ani Satz were active participants. At the Brown workshop, my official commentators were John Tomasi, David Estlund, Neera Badhwar, Corey Brettschneider, Peter Vallentyne, and Arthur Applbaum.

I thank Galina Bityukova of the Central Asian Resource Center in Almaty, Kazakhstan, for assembling twenty-one faculty from nine post-Soviet republics to spend a week discussing the book. Giancarlo Ibarguen and Manuel Ayau, president and past president, respectively, of Francisco Marroquin University, organized a two-week visit to Guatemala where I presented nine lectures to various audiences. Michael Smith, Geoff Brennan, and Bob Goodin arranged for me to spend ten weeks at the Research School for Social Sciences at Australian National University in 2002. Thanks also to Jeremy and Pam Shearmur for welcoming me into their home outside Canberra.

I thank Michael Pendlebury for arranging a three-week visit to the University of Witwatersrand in 1999 where I presented early versions of several of these chapters. I thank Horacio Spector and Guido Pincione for the opportunity to present much of this material during visits to Torcuato di Tella School of Law in Buenos Aires. I thank David and Laura Truncellito for helping to arrange lectures at Chen Chi University, National Chung Cheng University, and University of Taiwan, and for an unforgettable week touring the island. I thank the Centre for Applied Ethics and Green College at the University of British Columbia for their splendid hospitality in the spring semester of 2000 and likewise the Social Philosophy and Policy Center at Bowling Green State University in the fall of 1999. For single lectures, I wish to thank audiences and organizers at Michigan, Yale, UNC–Chapel Hill, Ohio, Rochester Institute of Technology, Santa Clara, Auckland, Alabama-Birmingham, Tulane, Georgetown, West Virginia, and James Madison.

I wish to thank all the wonderful people at the Liberty Fund in Indianapolis for their stunningly generous support in the aftermath of my surgery when I needed peace and quiet so I could learn how to think again. I thank the Earhart Foundation and Institute for Humane Studies for sustaining not only me but several of the University of Arizona's students over the years. Needless to say, my greatest debt is to the University of Arizona. It is home, and I thank my colleagues for making it feel that way. More than anyone else (which is saying a lot), Chris Maloney makes life in the department a thing of joy, from when I arrive in the morning until we walk home together in the evening.

All these good people, in addition to those mentioned above and those I'll credit more specifically in footnotes to follow, have done what they could to make this a better book: Scott Arnold, Lawrence Becker, Matt Bedke, Jeremy Bendik-Keymer, Jason Brennan, Gillian Brock, Chris Brown, Allen Buchanan, Tom Christiano, Andrew Jason Cohen, David Copp, Tyler Cowen, Peter Danielson, Jonathan Dancy, Stephen Darwall, Amitai Etzioni, James Fishkin, Ray Frey, Gerald Gaus, Allan Gibbard, Walter Glannon, Charles Goodman, Rob Gressis, Chris Griffin, Allen Habib, Rosalind Hursthouse, Jenann Ismael, Frances Kamm, Scott LaBarge, Jason Lesandrini, Loren Lomasky, Gerry Mackie, David Miller, Fred Miller, Chris Morris, Jan Narveson, Cara Nine, Guido Pincione, Steve Pink, Frances Fox Piven, Thomas Pogge, James Rachels, Peter Railton, Dan Russell, John T. Sanders, Steve Scalet, Daniel Shapiro, David Sobel, Horacio Spector, Christine Swanton, Mark Timmons, Mary Tjiattas, Kevin Vallier, David Velleman, Will Wilkinson, Elizabeth Willott, Matt Zwolinski, and those who toiled as anonymous readers for the Press.

Some of Part 2 previously appeared in "How To Deserve," *Political Theory*, 30 (2002) 774–99. DOI:10.1177/0090591702238203. © Sage Publications 2002. Some of Part 6 previously appeared in "History & Pattern," *Social Philosophy & Policy*, 22 (2005) 148–77. Part 4 incorporates material from "Equal Respect & Equal Shares," *Social Philosophy & Policy*, 19 (2002) 244–74. Chapter 22 updates and reworks material first covered in *Social Welfare and Individual Responsibility* (1998). © Cambridge University Press. An earlier version of Chapter 23 was published as "Diminishing Marginal Utility," *Journal of Value Inquiry*, 34 (2000) 263–72. © Kluwer Academic Publishers. Reprinted with kind permission of Springer Science and Business Media.

PART 1

WHAT IS JUSTICE?

1

The Neighborhood of Justice

Thesis: Theorists disagree. It is not their fault. Theorizing does not lead to consensus.

PRELIMINARY SURVEY

When I survey the terrain of justice, here is what I see. What we call justice is a constellation of somewhat related elements. I see a degree of integration and unity, but the integrity of justice is limited, more like the integrity of a neighborhood than of a building. A good neighborhood is functional, a place where people can live well. Yet, good neighborhoods are not *designed* in the comprehensive way that good buildings are. (Indeed, designed communities feel fake, like movie sets, with histories too obviously tracing back to the dated plan of a single mind.)

Is there a defining property of the neighborhood of justice, in virtue of which the word applies? Yes, Part 1 explains, but the property is general and formal; how it translates into more substantive principles depends on context. Parts 2 through 5 reflect on four substantive elements: desert, reciprocity, equality, and need. Part 6 pays homage to John Rawls and Robert Nozick, who "arguably framed the landscape of academic political philosophy in the last decades of the twentieth century."[1] My theorizing is inspired by (although perhaps only vaguely resembles) theirs.

[1] Fried 2005, 221.

THEORIZING

If justice is a neighborhood, then a *theory* of justice is a map of that neighborhood. The best theory will be incomplete, like a map whose author declines to speculate about unexplored avenues, knowing there is a truth of the matter yet leaving those parts of the map blank. A theory evolves toward representing the neighborhood more completely, in the hands of future residents who have more information and different purposes, even as the neighborhood itself changes.

I have become a pluralist, but there are many pluralisms. I focus not on concentric "spheres" of local, national, and international justice nor on how different cultures foster different intuitions, but on the variety of contexts we experience every day, calling in turn for principles of desert, reciprocity, equality, and need. I try to some extent to knit these four elements together, showing how they make room for each other and define each other's limits, but not at a cost of twisting them to make them appear to fit together better than they really do. Would a more elegant theory reduce the multiplicity of elements to one?

Would a monist theory be more useful? Would it even be simpler? The periodic table would in one sense be simpler if we posited only four elements – or one, for that matter – but would that make for better science? No. Astronomers once said planets *must* have circular orbits. When they finally accepted the reality of elliptical orbits, which have two focal points, their theories became simpler, more elegant, and more powerful. So, simplicity is a theoretical virtue, but when a phenomenon looks complex – when an orbit seems to have two foci, not one – the simplest explanation may be that it looks complex because it is. We may find a way of doing everything with a single element, but it would be mere dogma – the opposite of science – to assume we must.

ONLY THAT WHICH HAS NO HISTORY IS DEFINABLE[2]

Socrates famously wanted definitions, not merely an example or two, but in practice the way we actually learn is by example. Thus, I wonder: Does philosophical training lead us to exaggerate the importance of

[2] Nietzsche 1969, 80.

definitions? We do not need to know how to define 'dog' to know what a dog is. Why would justice be different?[3]

The project of analyzing 'dog' has not captured philosophical imaginations as analyzing justice has. But suppose only one of us will get tenure, and somehow the verdict turns on whether we classify jackals as dogs. The meaning of 'dog' suddenly becomes controversial. Those who fail to see it our way start to look unreasonable. Two lessons: First, we define and refine a concept's edges only when the need arises. Second, the needs spurring us to define the edges of justice tend to be conflicting. So, emotions tend to run high, exacerbated by the fact that rules of justice tell us not only what to expect from each other, but what to count as an *affront*. If injustice is an affront, not merely a disappointment, then theorizing about injustice will be hard. Strangely, if Joe's theory fails to condemn things we consider an affront, that in itself is a bit of an affront.

DISAGREEMENT

Reasonable people disagree about what is just. Why? This itself is an item over which reasonable people disagree. Our analyses of justice (like our analyses of knowledge, free will, meaning, and so on) all have counterexamples. We have looked so hard for so long. Why have we not found what we are looking for?

In part, the problem lies in the nature of theorizing itself. A truism in philosophy of science: For any set of data, an infinite number of theories will fit the facts. So, even if we agree on particular cases, we still, in all likelihood, disagree on how to pull those judgments together to form a theory. Theorizing per se does not produce consensus (although social pressure does).

Why not? Either an argument is sound, or not. So why isn't a theory compelling to all of us, if sound, or none of us, if not? My answer: Theories are not arguments, sound or otherwise. They are maps. Maps, even good

[3] For a superb concise discussion, see Gaus 2000, chap. 1. Gaus quotes Wittgenstein (§66) as follows:

> Consider for example the proceedings that we call games. I mean board-games, card-games, Olympic games, and so on. What is common to them all? Don't say; there *must* be something common, or that they would not be called games – but *look and see* whether there is anything common to all. For if you look at them you will not see something that is common to *all*, but similarities, relationships. And a whole series of them at that. To repeat: don't think, look!

maps, are not compelling. No map represents the *only* reasonable way of seeing the terrain. (Or at least, this is how I see it.)

We would be astounded if two cartography students separately assigned to map the same terrain came up with identical maps. We would doubt they were working independently. Theorists working independently likewise construct different theories. Not seeing how the terrain underdetermines the choices they make about how to map it, they assume their theory cannot be true unless rival theories are false, and seek to identify ways in which rival theories distort the terrain. Naturally, they find some, and such demonstration seems decisive to them, but not to rivals, who barely pay attention, preoccupied as they are with demonstrations of their own.

Although we disagree over theoretical matters, there is less discord over how we should treat each other day to day. I may believe, at least theoretically, that justice requires us to tear down existing institutions and rebuild society according to a grand vision. You may feel the same, except your grand vision is nothing like mine. Yet, when we leave the office, we deal with the world as it is. I find my car in the parking lot. You find yours. We drive off without incident. If we are to live in peace, we need a high level of consensus on a long and mostly inarticulate list of "dos" and "don'ts" that constitute the ordinary sense of injustice with which we navigate in our social world. The consensus we need to achieve concerns *how* (not why) to treat each other, and we need to achieve consensus where we do achieve it: in practice.

In effect, there are two ways to agree: We agree on what is correct, or on who has jurisdiction – who gets to decide. Freedom of religion took the latter form; we learned to be liberals in matters of religion, reaching consensus not on what to believe but on who gets to decide. So too with freedom of speech. Isn't it odd that our greatest successes in learning how to live together stem not from agreeing on what is correct but from agreeing to let people decide for themselves?

2

The Basic Concept

THESIS: Justice concerns what people are due. This much is uncontested, simply a matter of how we normally use the word. Exactly what people are due, though, cannot be settled entirely by conceptual analysis.

WHAT WE KNOW ABOUT THE BASIC CONCEPT

What is justice? It is a philosopher's question, and a philosopher might start by noting that when we ask what is justice, the term 'justice' is not a meaningless sound. We argue about justice, yet the very fact that we argue presupposes a level of mutual understanding. Because we share a language, we know we are not arguing about what is an eggplant, or what is the weather forecast, or what is the capital of Argentina. When we argue about justice, there may be much we do not know, but we know that justice has something to do with treating like cases alike.

We also know that treating like cases alike is not the whole of justice. Suppose a medieval king decrees that persons convicted of shoplifting shall have their left hand amputated. We protest. Such punishment is unjust! The king replies, "I don't play favorites. I treat like cases alike, so what's the problem?" Even if the king is telling the truth, this does not settle the matter. Amputating every thief's left hand is treating all alike, but evenhandedness (so to speak) is not enough. Impartiality is not enough. The idea of treating like cases alike is relevant, but there is more to justice than this.

Compare this to a second case. The king now decrees: Those found *innocent* of shoplifting shall have their left hand amputated. Again, we protest. Again, the king replies, "I treat like cases alike, so what's the problem?" What do we say now? In the first case, the king's conception of justice was barbaric. In the second, the king does not *have* a conception – not even a barbaric one. We know this because, if the king softens his stance and says from now on the innocent will merely be fined, not maimed, the punishment is no longer barbaric, but that does not fix the problem. The problem is, the king fails to grasp the concept. To argue about justice is to argue about what people are due.[4] Simply grasping the meanings of words tells us that punishment, even mild punishment, is not what innocent people are due.

While treating like cases alike does not rule out evenhandedly punishing the innocent, giving people their due does. When we ask "What is justice?" we make a decent start when we say, "Whatever else we may debate, justice is about what people are due." There is a limit to how far we can get by analyzing language, but we can get (and we just did get) somewhere.

We also know we can distinguish the basic *concept* from particular *conceptions* of what people are due. Thus, to John Rawls,

> it seems natural to think of the concept of justice as distinct from the various conceptions of justice and as being specified by the role which these different sets of principles, these different conceptions, have in common. Those who hold different conceptions of justice can, then, still agree that institutions are just when no arbitrary distinctions are made between persons in the assigning of basic rights and duties and when the rules determine a proper balance between competing claims to the advantages of social life.[5]

For present purposes, we do not need this much baggage. We need not take a stand on whether arbitrariness is always bad. (When we assign the right to vote in a given election, we arbitrarily distinguish between citizens celebrating their eighteenth birthday and citizens who are one day younger.) We also can leave open whether "competing claims to advantages of social life" are what need balancing. The basic concept is this: Normal conversation about doing justice to X is conversation about giving X its due. This shared concept is what enables us to propose different conceptions, then argue about their relative merits.

[4] I am not denying that we can do justice to animals, opportunities, and ourselves. Likewise, the Grand Canyon in some sense deserves its reputation. My focus here is on the connection between doing justice to X and giving X its due, not on what can substitute for the variable X.

[5] Rawls 1971, 5. See also Hart 1961, 155–9.

The idea that we can disagree about what justice requires presupposes that we agree that justice does, after all, *require.*

WHAT THE BASIC CONCEPT LEAVES OPEN

We know something about justice, then. The basic concept is not empty, since only so many things can count as a person's due. As noted, punishment cannot be an innocent person's due. Yet, if the concept is not empty, neither is it substantial enough to answer every question. For example, if Joe works harder than Jane, should Joe be paid more? What if Jane needs the money more than Joe does? Should Jane be paid more? The basic concept does not say. We cannot specify Jane's due simply by defining the term 'due.' How do we know when facts about how hard Joe works matter more than facts about how badly Jane needs the money?

Suppose, for argument's sake, that if Jane and Joe are equal in relevant respects, their employer ought to pay them equally. Now change the case slightly: Jane and Joe remain equal but have different employers. Must Joe's employer pay the same as Jane's? If Jane earns twenty-thousand dollars as a cook while Joe, a comparably good cook, earns thirty-thousand dollars at the restaurant next door, is that unjust? Do issues of justice arise when Jane and Joe are paid differently by the *same* employer, but not when their salaries are set independently by *different* employers? Why?

SEEKING A REFEREE

These questions suggest a problem. So long as rival conceptions are minimally credible (for example, so long as they do not endorse punishing the innocent), the basic concept will not have enough content to settle which is best. Neither can we settle anything by appealing to one of the rivals. Put it this way: If opposing players are disputing a rule, we cannot settle the dispute by consulting a player. We need a referee. We need to go beyond the kind of weight players have. We need a different kind of authority.

For example, we can choose a conception according to what sort of life that conception (institutionalizing, endorsing, acting on it) would help us lead.[6] This idea is not a conception of justice, and does not

[6] Williams (1985, 115) says this about conceptions of morality.

presuppose one, which means we can appeal to it without prejudice.[7] It can be a referee precisely because, on the field of justice, it is not one of the players.

The idea of being able to live well lacks the kind of *gravity* we associate with principles of justice. But since the idea is not a principle of justice, this is as it should be. After all, it is the players who inspire us, not the referees.

AMBIGUITY

We can flesh out the idea of living well in different, not necessarily compatible ways. Is the idea to meet basic needs, promote welfare in general, provide better opportunities, or foster excellence? In practice, and in the long run, such ends may all be promoted by the same policies. Even when the various standards are incompatible, though, they still matter. Asking whether a policy fosters excellence is not a mistake. Asking whether a policy empowers the least advantaged is not a mistake. Admitting that various things matter without always pointing in the same direction is not a mistake. If relevant standards sometimes point in different directions, that is life. Complexity and ambiguity are not theoretical artifacts.

JUSTICE: WHAT IS IT FOR?

Granting that the idea of living well is complex and ambiguous, the role justice plays in enabling us to live well may yet be (relatively!) simple and well defined. Suppose we do not see justice as a panacea; that is, suppose we accept that everyone getting their due does not guarantee that everyone is living well. Justice gives us something, not everything. What more specifically, then, is the point of justice? Here is a suggestion.

A *negative externality*, sometimes called a spillover cost, is the part of an action's cost that has an impact on bystanders.[8] Economists talk of internalizing externalities: that is, minimizing the extent to which innocent

7 Rawls says, "We cannot, in general, assess a conception of justice by its distributive role alone, however useful this role may be in identifying the concept of justice. We must take into account its wider connections; for even though justice has a certain priority, being the most important virtue of institutions, it is still true that, other things equal, one conception of justice is preferable to another when its broader consequences are more desirable" (1971, 6).

8 *Positive* externalities are benefits that spill over to enrich the lives of "innocent bystanders." The following discussion pertains more to negative externalities.

people are forced to bear the costs of other people's choices. If embracing a certain principle resolves a conflict, this is not enough to show that the principle is a principle of justice. However, if practicing a principle leads us to take responsibility for the consequences of our actions, then not only is it apt for resolving conflict; it also functions like a principle of justice, for it requires paying some attention to what people around us are due. Henry Shue says, "If whoever makes a mess receives the benefits and does not pay the costs, not only does he have no incentive to avoid making as many messes as he likes, but he is also unfair to whoever does pay the costs."[9] Externalities undermine harmony among parts of a polis, as per Plato. Our neighbors do not want to put up with drunk drivers, for example, and should not have to. To be just is to avoid, as best we can, leaving our neighbors to pay for our negligent choices.

I am not proposing an imperative to internalize externalities as a conception of, or even a principle of, justice. Instead, I am saying our reasons for wanting to limit the proliferation of negative externalities do not rest on any particular view of justice. Such reasons do not *derive* from a conception of justice but instead *support* any conception that leads people to internalize. Any theory of justice that would lead us away from internalizing negative externalities has an uphill climb toward plausibility. Internalizing negative externalities is only one aspect of what we need to live well, but it may be justice's characteristic way of helping us to live well. Justice is a framework for decreasing the cost of living together; the framework's larger point is to free us to focus less on self-defense and more on mutual advantage, and on opportunities to make the world a better place: that is, to generate positive rather than negative externalities.

This may not be the essence of justice. However, if what we call justice serves that purpose, then we have reason to respect what we call justice, and to be glad we have as much of it as we do.

If justice is itself foundational, it may have no deeper foundation. In that case, we can ask what justice is a foundation for. We can evaluate the soundness of a house's foundation without presuming there is something more foundational than the foundation. We ask what kind of life the house's occupants will be able to live, while realizing that foundations are not everything. Foundations facilitate the good life, but cannot guarantee it.

Later parts of this book do not rely overtly on this way of testing competing conceptions. This is partly because I wrote later parts first, partly

9 Shue 2002, 395.

because the test is nonstandard and accordingly controversial, and partly because my first aim is analytical: to assess how well the principles fare as conceptions of what people are due. When conceptual analysis is inconclusive, though, I step back to consider the point of seeing one thing rather than another as a person's due. In other words, if and when we cannot answer "What is justice?" head on, we can try an indirect approach, asking, "What kind of life goes with conceiving of justice in this way rather than that?" More precisely, we observe people and institutions, interpreting some people as reciprocating some laws as treating people equally, and so on, then ask whether that principal (reciprocity, equality), put into operation in that particular way (informing that action, relationship, philosophy, or institution) is helping. We do this while knowing that such interpretations are isolating only an aspect of what we observe, and may well be overemphasizing it.

We should keep in mind that the basic concept of justice often is determinate enough that we can see what is just without needing to appeal to other goals and values. For example, we know it is unjust to punish deliberately an innocent person. It is analytic that punishment is not what the innocent are due. We do not appeal to consequences to decide that. The only time we appeal to considerations external to the basic concept, such as consequences, is when the basic concept is not enough to sort out rival conceptions. That is all.

3

A Variety of Contestants

THESIS: Justice has several elements. No simple principle is right for every context.

ACCOUNTING FOR THE APPEARANCE OF PLURALISM

In a case of child neglect, we plausibly could say justice requires parents to tend to the child's needs. By contrast, if a century ago we had wondered whether women should be allowed to vote, it would have been beside the point to wonder whether women *need* to vote, because in that context what women were due was acknowledgement – not of their needs but of their equality as citizens. Talking as if justice is about meeting women's needs would have been to treat women as children. One way to account for such facts is to say different contexts call for different principles. Justice is about giving people their due; if we are not discussing what people are due, then we are not discussing justice. Yet, what people are due varies.

A MULTIPLICITY OF PRINCIPLES

Theories of justice typically are assembled from one or more of the following four elements. Principles of *equality* say people should be treated equally – providing equal opportunity, ensuring equal pay for equal work, and so on – or that people should have equal shares of whatever is being distributed.

Principles of *desert* say people ought to get what they deserve. People should be rewarded in proportion to how hard they work, or how much risk they bear in undertaking a given line of work, or how well they satisfy their customers. In a nutshell, principles of equality focus on what we have in common; principles of desert focus on how we distinguish ourselves.

Principles of *reciprocity* say that when Joe does me a favor, he puts me in debt. I now owe Joe a favor, not by virtue of what kind of person Joe is but by virtue of what kind of history we share. Again in a nutshell, where a principle of desert might focus on the character of a person, principles of reciprocity focus on the character of a relationship.

Finally, principles of *need* define a class of needs, then say a society is just only if such needs are met, so far as meeting them is humanly possible

PUZZLES

1. Almost everyone thinks justice has to do with equality. But equality along one dimension entails inequality along others. Whenever a politician proposes a tax cut, editorials appear saying 90% of the tax cut's benefit would go to the rich. The editorials never explain how this could be so. Suppose Jane Poor earns $10,000 and pays a flat 10%, while Joe Rich earns $100,000 and pays a flat 38%. Together they pay $39,000, 95% of which is paid by Joe Rich. If we cut both rates by one percent, Jane saves $100, while Joe saves $1,000, which is to say, Joe gets about 90% of the benefit. So, the pundits are right, although they never mention that Joe still pays $37,000, compared to Jane's $900, and of the $37,900 that Joe and Jane now are paying between them, Joe is still paying over 95% of that total. So, should inequality be reduced? *Which* inequality? The forty-fold difference in what Jane and Joe pay, or the seven-fold difference in what they have left after paying? How much inequality along one dimension can we tolerate for the sake of equality along another?

   Another puzzle comes from Rawls. Suppose, when people can profit from developing their unequal talents, everyone does better than they do under systems that flatten inequalities, flattening incentives in the process. In that case, prizing equality per se would seem irrational.
2. We think people ought to get what they deserve, but why think anyone deserves anything? We think we deserve credit for the excellence of our work, but not for what is mere luck. The puzzle, as Rawls notes: Our ability to work is itself mere luck; our social circumstances, our talents, and even our character are products of nature and nurture for which we can claim no credit. Therefore,

there is nothing for which credit is due, and the idea of desert is a mirage. True?

3. Most of us think justice has something to do with reciprocity. People who help us put us in their debt. Yet it is unclear when returning favors is a matter of justice. As Robert Nozick observes, people cannot put us in debt merely by conferring favors on us that we did not request and may not want.[10] Not only are there cases where justice does not require reciprocity; sometimes justice does not *permit* reciprocity. Karsten gave me my first academic job. Now, let us imagine, years later, Karsten applying for a job in my department. I know how to return the favor, but do I have a duty or even a right to take that into account when deciding how to vote?[11]
4. Most of us think justice has to do with need. Indeed, that justice has to do with need is part of the reason why justice matters as it does. Ordinarily, though, we see what people are due and what people need as different things. It is too simple to suppose X is Jane's due *simply* because Jane needs X. That is the wrong kind of connection. So, what other connection is there?

A more disturbing puzzle has to do with the fact that when we distribute according to X, we in effect reward people for supplying units of X. When we distribute according to X, we tend to get more X. This is a nice consequence when we distribute according to desert. What if the same were true of need: What if, when we distribute according to need, we tend to get more need? Obviously this is not merely a theoretical worry. Within your family, you want to make sure your children get what they need, but you do not want your children to think that the way for them to get your attention is to be needy. That would be a recipe for badly raised children.

What if we look outside the confines of the family? Suppose you visit Thailand. You want to give to children begging on the street, but your guide says the children were kidnapped from Cambodia and brought to Bangkok to beg. Every evening their captor feeds them if they've collected enough money, and cuts off a finger if they have not. (The threat of torture makes the children desperate, amputations make them look more pathetic, and it's all good for

[10] Nozick 1974, 93.

[11] Even if I could not vote for Karsten in that hypothetical case, it would remain true that I should do such things as mention his name in my book, so he knows I have not forgotten my duty to be worthy of the chance he gave me.

business.) It is as plain as a moral fact can be that those children desperately need your spare change. Yet if your guide is right, then if you distribute your money on the basis of need, you are financing an industry that *manufactures* need. So, there you are, needing to decide whether to give money to the child in front of you. What does justice have to do with need in that case? Why?

Later chapters revisit these puzzles, but offer no easy answers. I try to advance the conversation, not end it. I try to show why, despite puzzles, we are rightly reluctant to discard any of our basic categories: desert, reciprocity, equality, and need.

4

Contextual Functionalism

THESIS: The realms of justice, governed by different principles, are distinct, but sometimes clash.

A PLURALIST THEORY

I am wary of labels, but we can describe my theory as *contextual functionalism.* The theory is *pluralist* insofar as none of its four primary elements is an overarching standard to which the others reduce. The theory is *contextual* insofar as respective elements rule only over limited ranges.[12] Ranges are topics that are mutually exclusive *more or less,* jointly spanning the subject of justice *more or less.* Ranges are like tectonic plates insofar as their edges shift as our conception evolves. (Civil rights movements aim to extend the range of equality before the law.) The shifting can leave gaps in some places, and overlaps elsewhere. Thus, range-bound elements may leave some questions unanswered, and answer some questions in clashing ways. Moreover, places where principles clash are chaotic, insofar as "butterfly" effects – variations in detail – lead to different conclusions. So, is it unjust for me to hire my cousin? The details matter.[13]

[12] Christopher Wellman suggests my theory is like Walzer's in recognizing spheres of justice, but, Wellman also suggests, when Walzer speaks of spheres (1983, 28ff), he is seeing justice as relativized to forms of life within particular communities, whereas I speak of ranges of application of particular principles without assuming ranges are geographically limited. So, the metaphor of spheres suggests a similarity that is not there. Walzer does believe in a plurality of principles, so that similarity is real, but Walzer does not belabor this aspect of his theory. In any case, I will try not to exaggerate differences or similarities between my theory and the theories of others.

[13] I thank Clark Durant for the tectonic plate metaphor.

The theory is *functionalist* in saying we can try to resolve uncertainty over what to believe by asking what justice is for. There are considerations beyond justice. Some of them matter, regardless of whether they matter within the arena of justice. When considerations internal to the concept (for examples, analyzing the word 'due') do not settle which rival conception we should believe, we can ask what matters *outside* the arena, without prejudice to ideas that matter within. There is no assumption that what is outside the arena is more foundational than what is inside. The point is only that when we exhaust everything that matters inside the arena, without settling which conception of justice to regard as the real thing, we need not give up.

A CONTEXTUAL THEORY, CRUDELY STATED

Different principles apply in different contexts. A *context* is a question that motivates us to theorize. "What are my children due?" is one context. "What are my employees due (from me)?" is another. As we come to a map with a destination, so we come to a theory with a question, hoping for guidance. It is the topic of our pretheoretical question (children, employees, animals, and so on), not the theory per se, that specifies our theoretical context. In that sense, contexts are not theory laden.[14] So, here is a map of the neighborhood of justice. The topics are crude, specifying correspondingly crude contexts. We discuss refinements in a moment.

1. What are children due? They are due what they need.
2. What are citizens due? They are due equal treatment, that is, equality before the law.
3. What are partners due? They are due reciprocity.

[14] Gilbert Harman says, "There are no pure observations. Observations are always 'theory-laden.' What you perceive depends to some extent on the theory you hold, consciously or unconsciously. You see some children pour gasoline on a cat and ignite it. To really see that, you have to possess a great deal of knowledge . . . you see what you do because of the theories you hold. Change those theories and you would see something else." (1988, 120). The lesson to take from Harman's theory-laden "observation" is that "theory laden" is a relative term. Even the bare observation that the cat is on fire *can* be viewed as theory laden (depending on your theory), but it is less theory laden than a view that lighting the cat on fire is *wrong*, which in turn is less theory laden than a view that lighting the cat on fire is wrong *because* it causes needless suffering. My point: A *context* is a situation that raises a question like, "What do I owe the cat?" Answers will be theory laden, but the question itself, *relatively* speaking, is not.

4. What are contestants due? They are due fair acknowledgement of demonstrated merit.
5. What are employees due? They are due what they have earned.
6. Families at the twentieth income percentile correspond roughly to the class Rawls called "least advantaged." What are they due? As Rawls might have said, they are due maximum freedom compatible with similar freedom for all. They are due a chance to live in a society whose rising tide of prosperity does not leave whole classes behind. Their children deserve a chance to grow up in an open society, where humble origins are no great barrier to developing their full potential. Everyone deserves a chance, at least in a cosmic sense.[15]

HOW TO REFINE A CONTEXT: A CASE STUDY

In a pluralistic theory, the idea that people are due (for example) equal shares in one context is compatible with people being due something else in another context. Thus, the standard way of arguing by counter-example – constructing cases where equal shares would be monstrous – does not refute equal shares within a pluralistic theory. Instead, it does something more constructive: It shows us *when* a principle such as equal shares does not apply. It identifies limits.

Consider the first context listed earlier: questions about what children are due. A person of wisdom sees this as a crudely drawn context, so when she says, "Children are due what they need," she will not mean to be stating a universal law. She knows a full context is a nuanced thing, and any *verbal description* will be merely partial. So, she offers a general rule covering what she imagines to be a standard case. She realizes there will be counterexamples whose details go beyond what she meant to cover with her crude generalization. (Think of instruction manuals you have used while assembling a new piece of furniture. The task is simple, and you sincerely wish to understand the instructions, yet you still make mistakes. Is it any wonder that instructions for something vastly more complex – how to conceive of justice – could go astray in the hands of experts trained in the art of cleverly perverse interpretation?) So, asked what children are due, Jane says they are due what they need. Joe cleverly

[15] I speak of cosmic justice because saying what Jane is due leaves open whether anyone has a duty, or even a right, to make sure Jane gets her due.

replies, "What if my child is a grownup?" Jane hears Joe's counterexample not as refuting her answer, but as refining the original question. A true refutation shows that Jane's generalization is not true even in general.

This is what analytic philosophy is. If we could get past "philosophy to win," analytic philosophy would be a process of formulating generalizations for contexts that admit of further refinement. (I am, of course, generalizing.) We begin with something crude, something that would not be a good place to stop but that may be a good start. We can try to tear the proposal down, thoughtlessly, as a vandal would, or probe it with a view to discovering what might be built on it. Suppose Jane treats Joe's question as refining the original question. She answers in a fittingly refined way, saying: When I said parents ought to meet their children's needs – such is a child's due – I was imagining someone roughly six years old. You are asking about a context to which that answer does not apply. Here is my answer to your new question. Your adult children are also fellow citizens. Or if your adult child is also a business partner, or an employee, those refinements lead to different refinements of my answer. (People are more than one thing.)

Why would a young child's due differ from an adult child's due? Here is one answer. Sometimes, what your children need most is to be recognized and rewarded for meritorious performance. Or they might need you to establish and acknowledge a reciprocal relationship, such as when you pay them to mow the lawn. More generally, what your children eventually need is for you to start treating them like adults rather than like children.[16] Part of treating them like adults is treating them as having adult responsibilities. Treating them as having adult responsibilities involves, in part, acknowledging sharp limits to your obligation to meet their adult needs. It is part of the art of decent parenting: cutting children loose as they become able to handle the responsibility. There comes a point when distributing according to need is no longer what your children need. Your relationship to them is one context to which principles of justice apply, but context is not static. As children mature, the context evolves, gradually becoming a context to which different principles apply.

[16] John Locke (*Second Treatise*, chap. 6, sec. 55) says children are not born *in* a full state of equality, but they are born *to* it. I thank Chaim Katz for the reminder.

5

What Is Theory?

THESIS: Successful theories are maps, not attempts to specify necessary and sufficient conditions.

THEORIES ARE MAPS

Let us explore the idea that one way to see what a theory is, and what a theory can do, is to see a theory as a map.[17] We begin with a terrain (a subject matter), and with questions about that terrain. Our questions spur us to build theories – maps of the terrain – that articulate and systematize our answers. To know how to reach Detroit, we need one kind of map. To know how to be a good person, we need another map. Note: *Maps* do not tell us where we want to go.[18] Our questions predate our theorizing, and constitute our reasons to theorize in the first place.

Theories Are Abstractions

A map of Detroit is an artifact, an invention. So is a map of justice. In neither case does the terrain being mapped *really look like that.* A map of Detroit is stylized, abstract, and simplified. It otherwise would fail as a

[17] I thank Jenann Ismael for several educational and enjoyable conversations about theories as maps.

[18] This is equally true of scientific theorizing. For example, to those who want to understand nature in secular terms, Darwinism is a serviceable map. It does not explain everything, but it explains a lot. Darwinism is rejected by Creationists, though. Why? Not because it fails to help them understand the origin of species in secular terms, but because they have a different destination.

map. Yet a map can be accurate in the sense that it does not mislead. A given map will for some purposes have ample detail; for other purposes it will be oversimplified.

A map is not itself the reality. It is at best a serviceable representation. Moral theories likewise are more or less serviceable representations of a terrain. They cannot be more than that.

Fine Detail Is a Means to an End

When we construct a map, we leave out details that would merely confuse users. Fine detail is not an end in itself. We do not try to show current locations of every stalled car on the side of the road, and we do not call a map false when it omits such details. The question is whether users honestly wanting to follow directions would be led astray.

Comprehensive Scope Is a Means to an End

Existing theories tend to be like maps of the globe: a result of striving for comprehensive scope – for a principle or set of principles that covers everything. Real moral questions, though, often are more like questions about getting to campus from the airport. A map of the globe is impressive, but when we want to get to campus, the globe does not help. It is not even relevant.

Local maps do not say how to reach all destinations. Yet, though non-comprehensive, they almost always are what we want when we want a map. Why? Because they provide the detail we need for solving problems we actually have. The distant perspective from which we view the whole globe of morality is a perspective from which the surface looks smooth. Principles we stretch to cover the globe fail to make contact with the valleys of moral life. They do not help people on the ground to make moral decisions.

Theories Have Counterexamples

Typically, a counterexample's point is to show that a theory is not algorithmic: We could follow the letter of a theory and still arrive at the wrong destination. But we can consider it a folk theorem of analytic philosophy: *Any* theory simple enough to be useful has counterexamples. (This is a simple theory. Therefore, if correct, it has counterexamples.)

Counterexamples are warning signs, telling us that theories should not be trusted blindly, any more than a map should be trusted blindly in the face of road signs warning that the bridge ahead is washed out. Even simple travel instructions require interpretation, judgment, and experience. (Carbury said the turn was "about a mile." Have we gone too far? Is that the gas station he told us to watch for?) There is virtually no such thing as simply following instructions.

Theories Say What to Do in Context C, Not That We Are in Context C

Like it or not, we apply theories, not merely follow them. Put it this way: When we formulate *rules*, we try to formulate instructions that agents can follow, but when we formulate *principles* rather than rules, we are not even trying to formulate instructions that agents can simply follow. (There is comfort in the idea of following. It seems to relieve us of responsibility, whereas *applying* a theory requires good faith, wisdom, and experience, and leaves little room for doubt about who is choosing and who is responsible for the consequences.) Those who want principles of justice to be "idiot-proof" have the wrong idea about what a theory can do.

If your destination is the campus, a city map may tell you to turn left at First and Broadway, but by itself an ordinary map cannot tell you what to do right now unless you already know from experience and observation that you are at the corner of First and Broadway. An ordinary road map does not come with a red X saying, "You are here." Ordinary maps depend on a user to know where he or she is, and where he or she wants to go.

Theories are like ordinary maps in that respect. Even if a theory says unequivocally that principle P applies in context C, we still need to decide whether our current situation is enough like C to make P applicable. Unequivocal though principle P may be, we still need wisdom and experience to judge that the time for principle P has come.[19]

Different Destinations Call for Different Maps

Our purposes change. We seek answers to new questions, calling for a new map. A map of the city is for one purpose; a map of the solar system is for another. Likewise, a theory that maps a public official's duties may be quite different from a theory that maps a parent's duties.

[19] I owe the following thoughts to a conversation with Fred Miller: Whether a plastic model of the Parthenon is accurate has nothing to do with the fact that the model is made of plastic, because viewers somehow understand that the model is not representing the Parthenon as made of plastic. If the model were to depict the Parthenon as *circular*, that would make it false, because the model's shape is a *depiction* in a way that the plastic material is not.

Note: If we have more than one purpose, we may need more than one map *even if* there is only one ultimate reality.[20]

When Maps Overlap, They Can Disagree. So What?

Suppose I have two maps, and they disagree. I infer from one that I should take the freeway; the other says the freeway is closed. If I discard one, I make disagreement vanish, but that doesn't solve the problem. Disagreement is informative, telling me I need to pay attention. I cannot trust any map blindly. So, when maps are imperfect, there are worse things than having more than one. If I notice that they disagree, I check whether one of my maps is out of date, or consult a local resident. If I see grains of truth in incompatible theories, must I discard one for the sake of consistency?[21] No, not if theories are maps.

THEORIES ARE COMPROMISES

When we theorize, we seek to render what we know simple enough to be understood, stated, and applied. If we try to describe verbally every nuance of justice's complexity, we get something so unwieldy that it may not appear to be a theory at all. If instead we try to simplify, homing in on justice's essence, we get incompleteness or inaccuracy. The task is like trying to represent three-dimensional terrain in two dimensions. Mapmakers projecting from three dimensions onto two can accurately represent size or shape, but not both. Mercator projections depict lines of longitude as parallel, more or less accurately representing continental shapes at a cost of distorting relative size. Greenland looks as big as Africa but in fact is one fourteenth as large. Peters projections also treat lines of longitude as parallel, but solve "Greenland" problems by collapsing vertical space at polar latitudes. Relative sizes are reasonably accurate, but shapes are distorted. Goode's Homolosine is better at representing individual continents at a cost of depicting the world as a globe whose surface has been peeled like an orange.

In short, map making, like theorizing, is a messy activity. Mapmakers choose how to represent worlds, and there is no perfect way of

[20] My theory that theories are like maps is a theory: a way of systematizing and articulating how I see the activity of theorizing. The activity of theorizing is the reality; my "map theory" is my attempt to describe that reality. If my "map theory" is correct, it will have the limitations that maps tend to have.

[21] Robert Louden (1992, 8) says, "the existence of conflicting types of ethical theories is both intellectually healthy and close to inevitable."

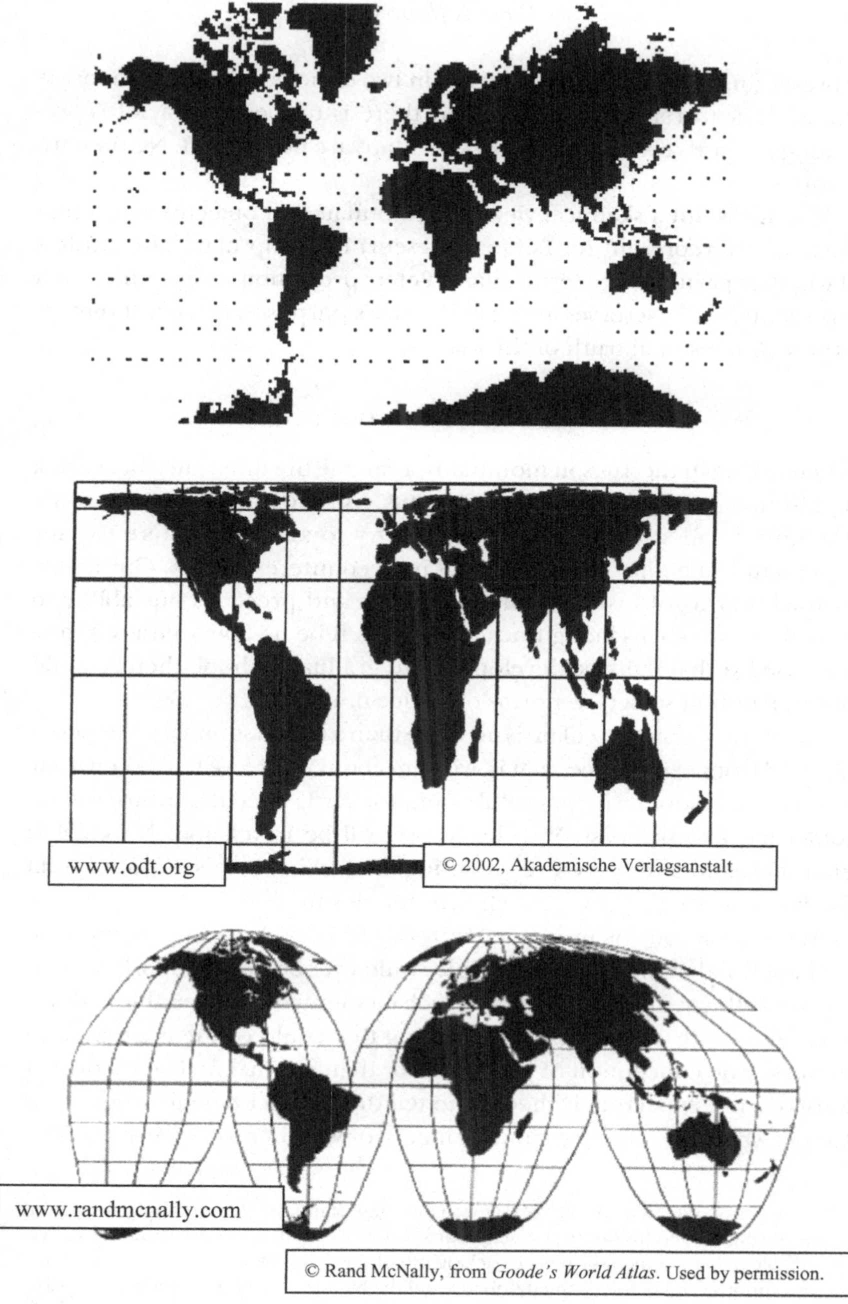

FIGURE 5.1 Mercator Projection, Peters World Map, Goode's Homolosine. Source: The Peters Projection World Map was produced with support of the United Nations Development Programme. For maps and related teaching materials contact: ODT, Inc., Box 134, Amherst MA 01004 USA; (800-736-1293; Fax: 413-549-3503; E-mail: odtstore@odt.org).

representing three-dimensional truth in two dimensions. Moral theorists choose how to represent justice, and there is no perfect way of representing in words everything we believe. Maps are not perfect. Neither are theories.

Yet, this is not a skeptical view! There remains an objective truth that the map can represent (or fail to represent) in a helpful way. Regardless of whether partisans of Mercator and Peters projections ever settle which representation best serves a particular user's purposes, there will remain a three-dimensional truth of the matter.

ARTICULATING THE CODE

When hiking in the Tucson mountains, I can see the difference between a pincushion cactus and a hedgehog cactus. I *see* the difference even while doubting I can *state* the difference. If I try to state the difference, my statement will be incomplete, or will have counterexamples. Our ability to track norms of justice similarly exceeds and precedes our ability to articulate the norms being tracked. Indeed, if being able to track X presupposed verbal skills we develop only in graduate school, then X could not function in society as norms of justice must.

Any code we can articulate is no more than a rough summary of wisdom gleaned from experience, that is, wisdom about where we have been. Our articulated wisdom will be useful going forward, since the future will be somewhat like the past. Yet, the future will be novel, too. No code is guaranteed to anticipate every contingency, which is to say, no formula (so far) unerringly prescribes choices for all situations.

We can list four or more elements of justice without ever being sure we have listed everything that people could ever be due. Similarly, we can list metaethical standards to which such elements are answerable without ever being sure we have listed everything that could count as a reason to endorse one conception of justice rather than another.[22] The theorists I know do not expect their theories to tell them what grade to assign, how to vote when the hiring committee meets, or whether to cancel class. The

[22] Legal reasoning often appeals to a "reasonable man" standard. Whether Bob is negligent for having backed his van over a neighbor's bicycle depends on what precautions a reasonable person would take before backing out of the driveway, and whether taking such precautions would have enabled Bob to avoid the bicycle. What is nice about reasonable person standards is that they do not raise false hopes regarding how comprehensive and how unified the enumeration of reasons that constitutes a theory can be. If Bob had to back his van through the neighbor's fence in order to run over the bicycle, Bob presumably is at fault. However, the fence's salience derives less from a list of principles than from grasping the details of the case.

wisdom and insight that enable us to see what to do are not precipitates of a theory in any straightforward way, although theorizing may contribute to their development.

Knowing which principle to apply requires judgment. Judgment is codifiable in a way, yet exercising judgment is not like following a code. Consider a simpler issue: Can a code tell investors when to buy and sell stocks? Market analysts look at histories of price fluctuations and see patterns. Patterns suggest formulas. Occasionally someone tries to sell such a formula, offering proof that the formula would have predicted every major price movement of the last fifty years. Investors buy the formula, which promptly fails to predict the next major move. My point: Many phenomena are codifiable – exhibiting a pattern that, after the fact, can be expressed as a formula – but that does not mean the formula will help us make the next decision.

So, when business majors in ethics courses ask for "the code" the following of which is guaranteed to render all their future business decisions beyond reproach, we may have little to say, even if we think such a code is, in principle, out there awaiting discovery. Business majors tend to understand stock markets well enough to know they can expect only so much from a stock-picking code. Responsibility for exercising judgment ultimately lies with them, not with any code. Some of them have not done enough moral philosophy to know they likewise can expect only so much from a moral code. But we can tell them the truth: Philosophers are in the business of articulating principles, not rules and not codes. Moral wisdom is less like knowing answers to test questions and more like simply being aware that the test has begun.[23]

I COULD BE WRONG

The periodic table is a theoretical structure, but is literally a map rather than an analysis. It also is, like my theory, thoroughly metaphorical, defining families of elements – alkali metals, noble gases – more or less according to how they behave. (My four elements turn out to be families: at least two kinds of deserving, three ways of responding to favors, and two

[23] Think of experiments in moral psychology where people fail to lend aid, fail to stand up for the truth, or succumb to pressure to torture fellow subjects. Now imagine the guy in the lab coat warning subjects that the point of the experiment is to test their moral integrity. My conjecture: Such warning would systematically affect subjects' behavior. Why? Not because the lab coat would be giving out *answers*. He would not be. All the lab coat would be doing is warning subjects that they are about to be tested. That life is about to test their character, though, is something people of wisdom get up every morning already knowing.

dimensions of equality, plus a complex hierarchy of needs.) And like my theory of justice, the periodic table is open ended, allowing for discovery or even invention of new elements. The table is a simple, elegant, fruitful way of organizing the information we have. It may even be the best way, but it is not *necessarily* so. If it is the best way of organizing the information we have, it need not remain so as new information comes in.

I have not tried to formulate necessary and sufficient conditions for X being just. There is only so much to gain from trying to articulate such conditions, and there are other kinds of analysis. (Economists tend to look not for necessary and sufficient conditions but for functional relationships: how Y varies as a function of X. A rise in the money supply is neither necessary nor sufficient for a rise in the inflation rate, but that is not the point. The point is that, other things equal, changing the money supply will affect prices.)

No philosopher is widely regarded as having succeeded in developing a viable theory of justice. I am under no illusion that mine will be the first. I do not represent any of this as compelling. Your way of understanding justice will differ from mine. You will have different answers, perhaps even different questions. That is not a problem.

I offer my results as meditations, not deductions. Gaps in a theory fire imaginations (or at least inspire replies), so I have not tried to hide the gaps. Socrates taught us that wisdom is not about how much we know; it is about seeing how much more there is to learn. Some aspects of this terrain remain hidden to me. The best I can do is to leave them alone until I learn more.

DISCUSSION

Are theories of justice of like road maps? One view: We evaluate a road map's accuracy by checking the terrain, whereas in moral philosophy, there is no terrain – no fact – out there to check. Another view: The facts *are* out there, facts about the kind of life we live if we (or our institutions) operate by one conception of justice rather than another. Of course, before a road map can point us in the right direction, we must decide where we want to go. We select a map to fit our destination, not the other way around. Are theories of justice in this way *too much* like road maps? Do we long for a treasure map, directing us to aim at a spot? We want our favorite reasons (to equalize, to reciprocate, and so on) to be more than mere reasons. We want our destination (the spot we choose to call justice) to be *compelling*. So, should we hold out for a treasure map, or settle for a road map?

PART 2

HOW TO DESERVE

6

Desert

I pulled over. The cop pulled in behind. Walked to my window, peered inside, asked for my license and registration.

"New in town?"

Yes, I said. Got in five minutes ago.

"Know what you did wrong?"

"Sorry. There was no stop sign or stop light. The cars on the cross street were stopped, so I kept going."

The cop shook his head. "In this town, sir, we distribute according to desert. Therefore, when motorists meet at an intersection, they stop to compare destinations and ascertain which of them is more worthy of having the right of way. If you attend our high school track meet tomorrow night, you'll see it's the same thing. Instead of awarding gold medals for running the fastest, we award them for giving the greatest effort. Anyway, that's why the other cars honked, because you didn't stop to compare destinations."

The cop paused, stared, silently.

"I'm sorry, Officer" I said at last. "I know you must be joking, but I'm afraid I don't get it."

"Justice isn't a joke, sir. I was going to let you off with a warning. Until you said that."

People ought to get what they deserve. And what we deserve can depend on effort, performance, or on excelling in competition, even when excellence is partly a function of our natural gifts.

Or so most people believe. Philosophers sometimes say otherwise. At least since Karl Marx complained about capitalist society extracting surplus value from workers, thereby failing to give workers what they deserve, classical liberal philosophers have worried that to treat justice as a matter of what people deserve is to license interference with liberty.

Rawls likewise rejected patterns imposed by principles of desert, calling it

> one of the fixed points of our considered judgments that no one deserves his place in the distribution of natural endowments, any more than one deserves one's initial starting place in society. The assertion that a man deserves the superior character that enables him to make the effort to cultivate his abilities is equally problematic; for his character depends in large part upon fortunate family and social circumstances for which he can claim no credit. The notion of desert seems not to apply to these cases.[1]

Rawls's view is, in a way, compelling. Inevitably, our efforts are aided by natural gifts, positional advantages, and sheer luck, so how much can we deserve? And if our very characters result from an interplay of these same factors, how can we (capitalists and proletariat workers alike) deserve anything at all?

Does Rawls leave no room for desert? Rawls's intent may have been narrower: simply to eliminate a rival to his difference principle (Chapter 31) as a test of the justness of basic structure. Whatever Rawls intended, though, his critique of desert has no such surgical precision. We know Rawls intended his *two principles* to apply only to society's basic structure, but his critique of desert is not similarly constrained, and cannot be constrained merely by *stipulating* or *intending* that it be so constrained. When Rawls says, "the concept of desert seems not to apply" to cases where outcomes are influenced by natural advantages or by character, he is implicating the larger moral universe, not merely basic structure. In particular, he wants to say the larger moral universe contains nothing (beyond his own first principle) to stop his difference principle from being *the* test of basic structure's justness. If Rawls's attack on desert is warranted, then the skepticism he is justifying is of a global nature.[2]

[1] Rawls 1971, 104. Rakowski (1991, 112) sees the passage as an "uncontroversial assertion, which even libertarians such as Nozick accept." Scheffler (1992, 307) likewise calls the passage "uncontroversial." Hayek (1960, 94) says, "A good mind or a fine voice, a beautiful face or a skilful hand, and a ready wit or an attractive personality are in large measure as independent of a person's efforts as the opportunities or experiences he has had." Hayek insists it is neither desirable nor practicable to ask basic structure to distribute according to desert. Gauthier (1986, 220) says, "We may agree with Rawls that no one deserves her natural capacities. Being the person one is, is not a matter of desert," although Gauthier doubts that this fact has normative implications.

[2] Rawls sometimes says he is arguing not against desert per se but only against desert as a preinstitutional notion. See Chapter 11. I thank Matt Bedke for the thought that, recalling

Samuel Scheffler says, "none of the most prominent contemporary versions of philosophical liberalism assigns a significant role to desert at the level of fundamental principle."[3] If so, I argue, the most prominent contemporary versions of philosophical liberalism are mistaken. In particular, there is an aspect of what we do to make ourselves deserving that, although it has not been discussed in the literature, plays a central role in everyday moral life, and for good reason.

my Chapter 4, we may treat Rawls's argument as a generalization that we cannot accept as is, but that admits further refinement.

3 Scheffler 1992, 301.

7

What Did I Do to Deserve This?

THESIS: Skeptics insist that, to be deserving, working hard is not enough; we must also deserve credit for being destined to work hard. While this skeptical theory is not incoherent, neither is there any reason to believe it.

THE "BIG BANG" THEORY

Nearly everyone would say people ought to get what they deserve. But if we ask *what* people deserve, or on what basis, people begin to disagree. A few will say we deserve things simply by virtue of being human, or being in need. Many will say we deserve reward for our efforts, or for the real value our efforts create. It is not necessary, and may not even be feasible, to produce a complete catalog of all possible desert bases. Suffice it to say, the standard bases on which persons commonly are said to be deserving include character, effort, and achievement.[4]

What are we doing when we judge someone to be deserving, that is, when we acknowledge someone's character, effort, or achievement? Here is a suggestion: to judge Bob deserving is to judge Bob worthy. It is to judge that Bob has features that make a given outcome Bob's just reward.[5] Intuitively, although less obviously, to acknowledge that there are things Bob can do to be deserving is to acknowledge that Bob is a

4 Feinberg (1970, 58) coins the term "desert base" to refer to what grounds desert claims. The idea is that well-formed desert-claims are three-place relations of the form "P deserves X by virtue of feature F."

5 See Sher 1987, 195. See also Narveson 1995, 50–1.

person: able to choose and to be responsible for his choices.[6] Something like this is implicit in normal deliberation about what a person deserves.

The skeptics' theory, in its most sweeping form, depicts desert in such a way that to deserve X, we must supply not only inputs standardly thought to ground a desert claim; we also must be deserving of everything about the world, including its history, that put us in a position to supply that input. In effect, the possibility of our being deserving ended with the Big Bang.

ALL ARE LUCKY; SOME ARE MERELY LUCKY

Recall Rawls's claim that a man's "character depends in large part upon fortunate family and social circumstances for which he can claim no credit."[7] Rawls repeatedly stressed, and thus evidently thought it relevant, that, "Even the willingness to make an effort, to try, and so to be deserving in the ordinary sense is itself dependent upon happy family and social circumstances."[8] In any case, many authors endorse such a view, and many are inspired to do so by Rawls.[9]

Needless to say, we all have whatever we have partly in virtue of luck, and luck is not a desert maker. Every outcome is influenced by factors that are morally arbitrary. ('Arbitrary' has a negative connotation, but without further argument, we are entitled only to say luck is morally neutral or inert. That is how I use the term here.) However, does the supposition that *some* of an outcome's causal inputs are arbitrary entail that all of them must be?

Of course not. Everyone is lucky to some degree, but the more a person supplies in terms of effort or excellence, the less weight we put on the inevitable element of luck. In any case, there is a big difference between being lucky and being merely lucky. The bare fact of being lucky is not what precludes being deserving. Being *merely* lucky is what precludes being deserving, because to say we are merely lucky is to say we have not supplied inputs (the effort, the excellence) that ground desert claims.

6 See Morris 1991.
7 Rawls 1971, 104.
8 1971, 74.
9 For example, Brock 1999. See her section on "How can we deserve anything since we don't deserve our asset bases?" Those who reject the premise (that, to be a desert maker, an input must itself be deserved in turn) include Narveson 1995, 67; Sher 1987, 24, and Zaitchik 1977, 373.

To rebut a desert claim in a given case, we must show that inputs that *can* ground desert claims are missing in that case. On a nonvacuous conception of desert, there will be inputs that a person can supply, and therefore can fail to supply. In general, finding that X falls outside a category is interesting only if falling *inside* is a real possibility.

A further point; there are infinitely many inputs that do not ground desert claims (luck, the Big Bang). So what? Skeptics say every causal chain has morally arbitrary links, but no one doubts that. The truly skeptical idea is that no chain has *non*arbitrary links. A skeptic says, "Even character, talent, and other internal features that constitute us as persons are arbitrary so long as they are products of chains of events containing arbitrary links. Every causal chain traces back to something arbitrary, namely the Big Bang. Therefore, nothing is deserved."

Some causal chains work their way through features internal to persons; it would be strangely credulous for a skeptic unquestioningly to assume this does not matter. If a so-called skeptic says, "Character is arbitrary," then someone who is properly skeptical replies, "Compared to what?" We distinguish outcomes that owe something to a person's character from outcomes that do not. Desert makers, if there are any, are relations between outcomes and internal features of persons. We need not (and normally do not) assume anything about what caused those features.

WHEN HISTORY DOESN'T MATTER

Is it odd that we normally make no assumptions about a desert maker's causal history? What if we had been talking about features of nonpersons? Joel Feinberg observes, "Art objects deserve admiration; problems deserve careful consideration; bills of legislation deserve to be passed."[10] John Kleinig says the Grand Canyon deserves its reputation.[11] Such remarks are offered as small digressions, noted and then set aside, but they point to something crucial. We *never* say the Grand Canyon deserves its reputation only if it in turn deserves the natural endowments on which its reputation is based. We *never* question artistic judgments by saying, "Even the greatest of paintings were caused to have the features we admire. Not one painting ever did anything to deserve

[10] Feinberg 1970, 55.
[11] Kleinig 1971.

being caused to have those features." Intuitively, obviously, it doesn't matter.

Skeptics assume it does matter in the case of persons, but the assumption is groundless. So far as I know, it has never been defended. As with nonpersons, when a person's internal features support desert claims, the support comes from appreciating what those features are, not from evidence that they are uncaused.

Some will say desert claims about paintings do not mean the same thing as desert claims about persons. Not so. The meaning is the same; what changes are the stakes. We do not need to reject claims about what paintings deserve to make room for our favorite principle of distributive justice; we need only reject claims about what *persons* deserve. This difference in what is at stake is why Big Bang theories are deployed only against desert claims made on behalf of persons. Stakes aside, though, Big Bang theories are as unmotivated for persons as for paintings.

BEING DESTINED TO WORK HARD

Here, then, is where matters currently stand. Ordinary thought about desert would be a recipe for skepticism if it were true that ordinary practice presupposes that people deserve credit for doing X only when people in turn deserve credit for having the ability and opportunity to do X. However, because ordinary practice assumes no such thing, ordinary practice has no such problem. We are left with two options. First, we can say no one deserves anything, and that is what we will say *if* we assume we deserve credit for working hard only if we in turn deserve credit for being "destined" to work hard. The second option is to say we deserve credit for working hard not because we deserve to have been destined to work hard, but simply because we did, after all, work hard. The latter is our ordinary practice.

Neither option is compelling. We are not forced to believe in desert; neither are we forced to be skeptics. We decide. We can ask whether we treat people more respectfully when we give them credit for what they do or when we deny them credit. Or we can ask what kind of life we have when we live by one conception rather than another. These are different questions, and not the only ones we could ask. Perhaps the answers all point in the same direction. Perhaps not. Sweeping skepticism is unattractive to most people, but there is no denying that skepticism is an option, and that some do choose to be skeptics.[12]

Refuting skeptics and answering, "How can we deserve anything at all?" are different tasks. We can answer the question, but not by refuting skeptics. For those who want an answer – who *want* an alternative to skepticism – my objective is to make room within a philosophically respectable theory of justice for the idea that there are things we can do to be deserving.

DO I DESERVE THIS?

When we consider how much sheer good luck we needed to get where we are today, it is natural for us to wonder, "Do I deserve this?" What does the question mean?

If we translate the question as, "What did I do, at the moment of the Big Bang, to deserve this?" the answer is, "Nothing. So what?" If we translate the question as, "What did I do, before being born, to deserve this?" the answer again is, "Nothing. So what?" However, if we translate the question as "What did I *do* to deserve this?" then the question will have a real answer. Also eminently sensible would be to ask, "What *can* I do to deserve this?" This question too will have an answer. The answer may be that, as it happens, there is nothing I can do, but that is not preordained. A theory that lets us ask and answer this question is a theory that lets the concept of desert be what it needs to be in human affairs: a message of hope that is at the same time life's greatest moral challenge. Such a theory acknowledges the existence of persons: beings who make choices and who are accountable for the choices they make.

In summary, a genuine theory of desert tells us what to look for when investigating what particular people have done. A genuine theory will not say what "Big Bang" theories say: namely, we need not investigate actual

[12] Walzer (1983, 260) says, "Advocates of equality have often felt compelled to deny the reality of desert." In a footnote, Walzer says he is thinking of Rawls. Walzer sees Rawls's argument as supposing "the capacity to make an effort or to endure pain is, like all their other capacities, only the arbitrary gift of nature or nurture. But this is an odd argument, for while its purpose is to leave us with persons of equal entitlement, it is hard to see that it leaves us with *persons* at all. How are we to conceive of these men and women once we have come to view their capacities and achievements as accidental accessories, like hats and coats they just happen to be wearing? How, indeed, are they to conceive of themselves?"

histories of particular people, since we know a priori that no one deserves anything.

PUZZLES

1. We entertained the idea that character is an accident of nature/nurture for which we deserve no credit. In some way, that must be true, but where does it end? Could I have had an altogether different character, or is there a point beyond which such a person would not have been me? Am I lucky I was born human when I could have been a seagull? (Is there a seagull out there that could have been me?) Would we be wrong to say luck is a matter of what happens to me, whereas my basic nature (the fact that I have my character rather than yours) did not *happen* to me – it *is* me?
2. We argue about whether we have free will, but not about whether we are capable of feeling pain. Why?[13] Is one harder to prove than the other? Are we more in need of proof in one case than the other? Is that generally how we learn that something is true – by *proving* it? Is that how I learn that there will be a sunrise tomorrow morning, or that some of my friends believe every event has a cause? (If we were to solve the mystery of how consciousness arose in what had been a merely material world, I expect there would be no residual mystery concerning free will.)

[13] If free will were an on/off switch that happens to be turned on for beings like ourselves, then wanting to have free will would be pointless, like wanting to be composed of atoms. A freedom worth *wanting* will be a kind that can be *at stake*, that can be gained or lost. To some extent, I think, our free will indeed is like this, and not simply an on/off switch. I have seen enough of contemporary psychology to accept that unity of consciousness, and the free will that goes with it (1) are achievements, not givens, and (2) are achieved in degrees. Moreover, (3) how conscious and how free our minds are depends to some extent on how free our institutions (especially our classrooms) are.

8

Deserving a Chance

Thesis: There is more than one way to be deserving, and in particular, more than one way to deserve an opportunity. We sometimes deserve X on the basis of what we do after receiving X rather than what we do before.

HOW DO I DESERVE THIS?

Suppose we know *what* a person has to do to be deserving. Is there also a question about *when* a person has to do it? James Rachels says, "What people deserve always depends on what they have done in the past."[14] David Miller says, "desert judgments are justified on the basis of past and present facts about individuals, never on the basis of states of affairs to be created in the future."[15] Joel Feinberg says, "If a person is deserving of some sort of treatment, he must, necessarily, be so in virtue of some possessed characteristic or *prior* activity."[16]

If we are not careful, we could interpret such statements in a way that would overlook an important, perhaps the most important, kind of desert-making relation. It is conventional that what we deserve depends on what we do, and that we deserve no credit for what we do until we do it. There may be a further aspect to academic convention, though, namely that when we first receive (for example) our natural and positional advantages, if we have not *already* done something to deserve them, it is too late.

[14] Rachels 1997, 176.
[15] David Miller 1976, 93.
[16] Feinberg 1970, 48. Emphasis added.

We are born into our advantages by mere luck, and that which comes to us by mere luck can never be deserved.

This further aspect is what I reject. I said being merely lucky precludes being deserving. I did not say, and do not believe, that being merely lucky at t_1 precludes being deserving at t_2. In particular, we do not deserve our natural gifts *at the moment of our birth*, but that need not matter. What matters, if anything at all matters, is what we do after the fact.[17] Let me make a claim that may at first seem counterintuitive:

We sometimes deserve X on the basis of what we do after receiving X.

Upon receiving a surprisingly good job offer, a new employee vows to work hard to deserve it. No one ever thinks the vow is paradoxical. No one takes the employee aside and says, "Relax. There's nothing you can do. Only the past is relevant." But unless such everyday vows are misguided, we can deserve X on the basis of what we do after receiving X.

How can this be? Isn't it a brute fact that when we ask whether a person deserves X, we look backward, not forward? If we concede for argument's sake that we look back, we would still need to ask: back from where? Perhaps we look back from where we are, yet mistakenly assume we look back from where the *recipient* was at the moment of receiving X. If we look back, a year after hiring Jane, wondering whether she deserved the chance, what do we ask? We ask what she *did* with it. When we do that, we *are* looking back even while looking at what happened after she received X. From that perspective, we see we can be deserving of opportunities.[18] We deserve them by not wasting them – by giving them their due, as it were.[19]

[17] In passing, there are desert bases that do not require action, such as when we say the Grand Canyon deserves its reputation. It deserves its reputation because of what it is, not because of what it did. I thank Neera Badhwar for noting the implication: Being merely lucky only *sometimes* precludes being deserving.

[18] I speak interchangeably of deserving a chance, being deserving of a chance, and being worthy of it. Sometimes, it is more natural to describe a person as being deserving of X rather than as deserving X, especially when the question concerns opportunity. But this is a verbal point. If a student said, "No one deserves anything, yet there is much of which people are deserving," we would think the student was making an obscure joke.

[19] Is this a sufficient condition? No. If something is wrong with the opportunity, as when we have a chance to use stolen property, then not wasting the opportunity does not suffice to show we deserve it. We could say the same of standard theories about deserving rewards: When we know the reward is stolen property, qualifying for it does not suffice to show we deserve it. In the same way, we may think we establish title to a previously unowned good by mixing our labor with it, without thinking that labor mixing can give us title to what otherwise is someone else's property.

Therefore, even if we necessarily look back when evaluating desert claims, the point remains that the use sometimes – even when the use occurs after the fact – bears on whether a person was worthy of the opportunity. Imagine another case. Two students receive scholarships. One works hard and gets excellent grades. The other parties her way through her first year before finally being expelled for cheating. Does their conduct tell us *nothing* about which was more deserving of a scholarship?

Can we save the convention (that whether we deserve X depends entirely on what happens before we receive X) by saying the students' conduct is relevant only because it reveals what they were like before the award? No. When we look back at the expelled student's disgraceful year, our reason for saying she did not deserve her award has nothing to do with speculation about what she did in high school. Both students may have been qualified for scholarships qua *reward.* Or equally *un*qualified: Suppose both were chosen via clerical error, and prior to that were equally destined for a lifetime of failure. The difference is subsequent performance, not prior qualification. What grounds our conviction that one is more worthy of the scholarship qua *opportunity* is that one student gave the opportunity its due; the other did not. Again:

We sometimes deserve X on the basis of what we do after receiving X.

TWO WAYS TO BALANCE THE SCALE

Needless to say, skeptics greet this conclusion with skepticism. Why? Part of the answer is that we learn as philosophers to focus on desert as a *compensatory* notion. The idea is, desert makers we supply before getting X put a moral scale out of balance, and our getting X rebalances the scale. To those who see desert as necessarily a compensatory notion, we deserve X only if X represents a restoring of moral balance. We deserve X only if we deserve it qua reward – only if our receiving X settles an account.

In ordinary use, though, desert sometimes is a *promissory* notion. Sometimes our receiving X is what puts the moral scale out of balance, and our subsequently proving ourselves worthy of X is what restores it. X need not be compensation for already having supplied the requisite desert makers. Sometimes it is the other way around. There are times when supplying desert makers is what settles the account.

In either case, two things happen, and the second settles the account. In compensatory cases, desert-making inputs are supplied first, then a reward settles the account. In promissory cases, an opportunity is given first, then supplying desert-making inputs settles the account. In promissory cases, a new employee who vows, "I will do justice to this opportunity. I will show you I deserve it" is not saying future events will retroactively cause her receiving X to count as settling an account *now*. Instead, she is saying future events *will* settle the account. Her claim is not that she is getting what she already paid for but that she is getting what she *will* pay for.[20]

So why does James Rachels assert that, "What people deserve always depends on what they have done in the past"?[21] Rachels says, "[T]he explanation of why past actions are the only bases of desert connects with the fact that if people were never responsible for their own conduct – if strict determinism were true – no one would ever deserve anything."[22] Crucially, when he says, "past actions are the only bases of desert," Rachels is stressing "actions," not "past." What Rachels sees as the unacceptable alternative is not a theory like mine, but rather the view that people deserve to be rewarded for having natural endowments. He is thinking of past actions versus past nonactions, and is not considering whether actions postdating X's receipt might be relevant. That is why Rachels could see himself as explaining why "past actions are the only bases of desert" when he says, "[I]f people were never responsible for their own conduct, . . . no one would ever deserve anything." Notice: this argument in no way connects desert bases to events predating X's receipt. The argument connects desert to action, but not particularly to *past* action.[23]

[20] Feldman (1995, 70–1) argues that a soldier who volunteers for a suicide mission can deserve a medal in advance. Perhaps, but see Chapter 9. In any case, Feldman's case is still an example of deserving a reward, not an opportunity. (Feldman does not claim people deserve opportunities.)

Jeremy Waldron and Fred Miller see forward-looking elements in Aristotle's discussion of meritocracy in distributing political offices. Aristotle (*Politics*, Book III, Chap. 12, 1282b, line 30ff) says, "When a number of flute players are equal in their art, there is no reason why those of them who are better born should have better flutes given to them; for they will not play any better on the flute, and the superior instrument should be reserved for him who is the superior artist." See Fred D. Miller 2001. Intriguingly, Waldron suggests a school might choose among candidates by comparing how meritorious the *school* would be if it hired one rather than another. See Waldron 1995, 573.

[21] Rachels 1997, 176.

[22] Rachels 1997, 180.

[23] An important caveat: Although Rachels and David Miller (1976) say what we deserve depends on what we did in the past, and never on the future, it would be anachronistic

Rachels also says, "People do not deserve things on account of their willingness to work, but only on account of their actually having worked."[24] There are reasons for saying this, and Rachels may be right when speaking of rewards. It may be analytic that *rewards* respond to past performance. However, rewards are not the only kind of thing that can be deserved. We sometimes have reason to say, "She deserves a chance." We may say a young job candidate deserves a chance not because of work already done but because she is plainly a talented, well-meaning person who wants the job and who will throw herself into it if given the chance.

A more senior internal candidate may be deserving in a different way: that is, worthy of reward for past performance. Yet, the idea that an inexperienced candidate can deserve a chance, for the reasons mentioned, is something most people find compelling. We can be glad they do, too, insofar as thinking this way leads them to give opportunities to people who are worthy in the promissory sense, that is, people who, when given a chance, give the opportunity its due.[25]

If we say a job candidate deserves a chance, and then, far from throwing herself into the job, she treats it with contempt, that would make us wrong. The promissory aspect of desert will have failed to materialize. She had a chance to balance the account and failed. If she treats the job with contempt, then she supplies neither the performance nor even the good faith effort that the hiring committee expected.

If instead the candidate fails through no fault of her own, then we cannot hold it against her. And if her failure is simply a stroke of unforeseeable bad luck, then neither can the committee blame itself for having chosen wrongly. They may say in retrospect that although the new employee failed to do justice to the opportunity, it was because she did not really get the opportunity the committee intended. By analogy, suppose we intend to give salt a chance to dissolve in water, but what we actually end up doing is giving salt a chance to dissolve in olive oil. If the salt fails to dissolve, we still insist the salt would have dissolved in water, given the chance.

to interpret them as rejecting my proposition that we can deserve X by virtue of what we do after receiving X. At the time, it had not yet occurred to anyone to be for, or against, my proposition.

24 Rachels 1997, 185.

25 Not all true statements about what we deserve have the status of desert *claims*. Claims in the relevant sense imply correlative duties, such as the duty to give claimants what they deserve. Someone who says Jane did justice to her opportunity may be expressing a truth without meaning to be making a claim on Jane's behalf against anyone else.

The possibility of bad luck notwithstanding, the fact remains that we sort out applicants for a reason. Typically, the point is not to reward someone for past conduct but to get someone who can do the job. That is why, by the time we reach t_2, the question is not what she did before the opportunity but what she did with it. The question at t_2 need not and typically does not turn on what was already settled at t_1.

A note on examples. Realistic examples are complex, raising issues beyond those intended by the theorist who brings them up. In this case, real-world hiring committees must juggle several criteria, not all of them having to do with desert. Some points might be better illustrated by speaking of tenure and promotion committees, where decisions are more purely a matter of desert but where candidates have enough of a history that it is harder to sort out backward versus forward-looking grounds for judging whether a candidate is deserving. Candidates often see their case as purely backward looking, but tenure committees do not. Tenure committees want to know that a candidate will not become deadwood – that past performance was not spurred mainly by a prospect of tenure qua *reward*. They want to be able to look back years later and say the candidate deserved tenure qua *opportunity*.

PUNISHMENT

Would I entertain a promissory theory of punishment? ("He may be innocent now, but if we put him in jail, he'll become the sort of person who belongs in jail.") No. We may view reward and punishment as two sides of the same compensatory coin, but there is no such parallel between opportunity and punishment. The transformative role of expectations (the fact that we tend to live up to them, or down, as the case may be) can justify the faith involved in granting an opportunity but cannot justify punishment.[26] If Jean Valjean wrongly is imprisoned and says, "OK, if they treat me like a criminal, I'll act like one," this does not vindicate the wrongful punishment. Indeed, the fact that the punishment induces punishment-worthy behavior further condemns the punishment. By contrast,

[26] George Rainbolt suggests that my promissory model may have a greater range of explanatory power than I give it credit for. In particular, if we have good reason to believe a prisoner convicted of a violent crime is unrepentant and indeed intent on repeating his crime upon being paroled, that is a reason for not granting parole. It is a reason not only in the sense that society has a right to protect itself from a confirmed and unreformed violent criminal, but also in the sense that the prisoner is undeserving of parole. Thus, the promissory model may underwrite some aspects of punishment after all.

if Valjean later is rocked by a bishop's kindness and says, "OK, if they treat me like a decent human being, I'll act like one," that *does* vindicate the bishop's kindness.[27]

Philosophical discussion of desert has often presupposed a bipartite model: That is, two things can be deserved: reward and punishment. But ordinary thought follows a tripartite model. We can deserve reward or punishment, to be sure, but we also can deserve a chance. We could reduce deserving a chance (or punishment, for that matter) to a species of deserving a reward. But it would be pointless, a spurious conceptual parsimony yielding no insight. A tripartite model is better at helping us to appreciate the nature and wisdom of ordinary moral practice.

REFINING THE PROMISSORY MODEL

To further clarify the nature of the promissory model, we should separate it into two elements. The first explains what we can say about Jane from the perspective of t_2. The second explains what we can say about her from the perspective of t_1.

Element (a): A person who receives opportunity X at t_1 can be deserving at t_2 because of what she did when given a chance.

Element (b): A person who receives opportunity X at t_1 can be deserving at t_1 because of what she will do if given the chance.

What does element (a) tell us? Element (a) tells us it can be true at t_2 that the account has been settled. Jane supplied inputs that did justice to X. We need not suppose Jane supplied those inputs at t_1. When we call Jane deserving at t_2, as per element (a), we are not denying that she may have been merely lucky at t_1. All we are saying is, when Jane got the chance to prove herself worthy, she did so.

Element (a) concerns what Jane can do to be deserving at t_2 even if she was merely lucky at t_1. By contrast, element (b) concerns how Jane can deserve X at t_1 not as a reward for past performance but as an opportunity to perform in the future. In other words, element (b) concerns how a committee nonarbitrarily could select Jane in preference to some other candidate. Jane is choiceworthy if she is the sort of person who will do justice to the opportunity. She may be choiceworthy in virtue of past performance, but the committee is not trying to *reward* past performance.

[27] Jean Valjean is a character from Victor Hugo's novel, *Les Misérables*.

They are trying to decide whether to count Jane's past performance as *evidence* that she will do justice to opportunity X – evidence that she *will* settle the account, given the chance.

There are various ways to formulate element (b). None are perfect. When we think of contexts like hiring decisions, it is natural to say a hiring committee is looking not merely for someone who theoretically can do the job, but for someone who *will* do the job given a chance, meaning she will do the job if we offer it to her, if she takes it, if there is no unforeseen catastrophe, and so on. Our invocation of element (b) at t_1 is, in effect, a prediction that by the time we get to t_2, we will be in a position to invoke element (a). We are predicting that by t_2, Jane will have supplied the relevant desert-making inputs. However, we are not merely wagering on future performance. Rather, we are wagering that Jane has desert-making internal features that will translate into future performance barring unexpected misfortune. We are saying she is the kind of person who will do the job given the chance.[28]

Element (a) says that although desert requires a balance between what Jane gives and what Jane is given, Jane need not move first. Element (b) says Jane can deserve opportunity X (in the sense of being choiceworthy) before she does her part. Element (a), by contrast, pointedly does not say Jane can deserve X before doing her part. Element (a) stresses that *even if* Jane deserves X only *after* doing her part, it *still* does not follow that she has to do her part before receiving X.

Element (a) therefore is the essence of the promissory model's departure from the idea that we deserve X only if we deserve it as a reward for past performance. So far as our purpose is to challenge this idea, we do not need element (b). We need some version of element (b) only insofar as we seek to vindicate ordinary practice – in particular our tendency to speak of candidates as deserving a chance by virtue of what they can and will do if we give them a chance.[29]

[28] When we think a machine will perform well if we give it a chance, we do not say the machine deserves a chance. We may say, "It is worth a try," but we do not mean the same thing when speaking of a person's character as when speaking of a machine's characteristics. I owe this point to Michael Smith.

[29] David Miller comes as close as any philosopher ever has, to my knowledge, to endorsing element (b). Miller says there are insuperable obstacles to interpreting *jobs* as rewards for past conduct (1999a, 159). When we say someone deserves a prize, we standardly base our judgement on past or present performance, but when we are making hiring decisions, the best qualified candidate, the one who deserves it, is the one who will perform it best, other things being equal (162). And "in the case of jobs past performance matters only as a source of evidence about a person's present qualities" (170).

A PUZZLE ABOUT PREDICTION

I said there are various ways of formulating element (b) and none are perfect. In spelling out element (b), we could interpret choiceworthiness as a question of either what is *true* of the candidate or what the committee *justifiably believes* about the candidate.[30] There are pros and cons either way. We may sometimes have reason to distinguish *evidence* that Jane will do well from the *fact* (if and when it becomes a fact) that Jane will do well. What makes Jane choiceworthy in the metaphysical rather than the epistemological sense is the fact that she truly is the kind of person who (barring unforeseen catastrophe) would supply the requisite desert makers and thus *become* deserving at t_2 in the sense of element (a).

If a committee concludes that Jane is choiceworthy at t_1, then whether the committee judged correctly (that is, whether it truly picked the right person, as opposed to whether they were justified in believing they picked the right person) remains to be seen. Is this a puzzle? If so, it is less a puzzle about desert and more a puzzle about prediction in general. Suppose at t_1 we say Jane will be married at t_2. Jane then gets married. In that case, events at t_2 have indeed settled the truth value of a claim uttered at t_1. Does anyone find this puzzling? The future event does not backward cause the prediction to be true; it simply settles that the prediction was true. Events at t_2 can settle the truth value of a claim such as, "She'll get married, given a chance." They also can settle the truth value of a claim such as, "She'll do justice to X, given a chance." There comes a time when we can say, "You said she'd get married; it turns out you were right," or when a committee can say, "We said she'd do justice to the opportunity; it turns out we were right." In either case, Jane settles what had been unsettled. Saying "she deserves X," meaning she will do justice to X given a chance, is no odder than saying "salt is soluble," meaning it will dissolve in water given a chance.

Insofar as the idea that Jane deserves a chance at t_1 depends on whether Jane has relevant dispositional properties at t_1, and insofar as a test of this idea lies in the future, element (b) implies that life sometimes involves decision making under uncertainty. Hiring committees judge which candidate is most worthy, with no guarantee that they are judging correctly.

[30] Recall David Miller's (1976) claim that "desert judgments are justified on the basis of past and present facts about individuals." I can agree that the epistemological *justification* of desert claims is backward looking, because that is where the information is, while still holding that *truth makers* for some desert claims can lie in the future. (We would say the same of predictions in general.)

When a committee judges at t_1 that Jane deserves a chance, they are placing a bet. They are judging her character. They may even transform her character, insofar as their trust may inspire Jane to become the kind of person they judge her to be. At t_1, though, it remains to be seen whether Jane is or will become that kind of person. Jane settles that later, in an epistemological sense, and perhaps in a metaphysical sense too, insofar as Jane will have to *decide*, not merely reveal, whether she really is that trustworthy, that hardworking, and so on. The committee will have to wait and see. Since life truly is difficult in this way, we can be glad to have a theory that correctly depicts the difficulty – that does not make life look simpler than it is.[31]

PEOPLE WHO NEVER GET A CHANCE

What can the promissory model say about unsuccessful candidates, or more generally about people who lack opportunity? What if there are more deserving candidates than positions for them to fill? Element (a) is silent on questions about people who never get a chance, but element (b) can say about unsuccessful candidates roughly what it says about successful ones; namely, they may deserve X insofar as they too would have done justice to X, given a chance. *My theory does not say people who lack opportunities are undeserving.*

According to my theory, there is something slightly misleading, or at best incomplete, in assessing a society by asking whether people get what they deserve. If desert matters, then often a better question is, do people *do something to deserve* what they get? Do opportunities go to people who will do something to be worthy of them?

My purpose here is to make room within a credible theory of justice for the idea that there are things we can do to be deserving. Specifically, we can deserve an opportunity. Moreover, whether we deserved an opportunity can depend on what we did with it. First, there are things we can do after the fact to balance the scale, making it fitting in retrospect that we got a chance to prove ourselves at t_1. Second, we can be choiceworthy even at t_1 insofar as a committee can see (or insofar as it is true) that we will do justice to the opportunity. The latter is not the core of my theory of desert, but it is a way of pushing the envelope and making sense of a central part of ordinary life.

[31] I thank Guido Pincione and Martín Farrell for their insight on this point.

9

Deserving and Earning

THESIS: The terms 'deserving' and 'earning' are nearly interchangeable in ordinary use, but there is a difference that matters when we speak of deserving opportunities.

EARNING

We commonly show our respect for what a person has achieved by saying, "You deserved it" or "You earned it." The words 'deserving' and 'earning' are nearly interchangeable in ordinary use. There is a difference, though, and it will be useful to give the difference a bit more emphasis than it gets in ordinary use.

A paycheck is not earned until the work is done. Upon being hired, I will do what I need to do to earn the paycheck, but the future does not settle that I have earned the paycheck *now*. I have not earned it until I do the work. Thus, while we do speak of people as deserving a chance even before they supply the requisite inputs, we do not speak of people as having earned a paycheck prior to supplying requisite inputs. Perhaps this is because what Jane deserves has more to do with her character, whereas what Jane has earned has more to do with her work. Jane's character can be manifest before she supplies the requisite inputs. Her work cannot similarly be manifest prior to supplying requisite inputs, since her work *is* the requisite input when the question concerns what she has earned. Jane can be deserving at t_1 in virtue of what she will do, if given a chance. To have earned a paycheck at t_1, though, she has to have done the work at t_1. Therefore, that she *would* earn the check at t_2 is not relevant to what Jane has earned at t_1, even though – according to element

(b) – it is relevant to whether Jane deserves a chance at t_1. Therefore, at t_1, the promissory model does not work for earning. There is no analog of element (b).

Somewhat to my surprise, though, there is an analog of element (a). We acknowledged that I have not earned the paycheck until I do the work. Does that mean I can earn the check only if I do the work *first*, before the check is issued?

No! In everyday life, we do not doubt that a new but trusted employee, paid in advance, can earn the money after the fact. Money is paid at t_1, then what was not true at t_1 becomes true at t_2: Namely, the scale is now balanced and money given at t_1 has been earned. It becomes true at t_2 that Jane did what she was paid to do.

Therefore, we cannot save the academic convention that desert is a purely compensatory notion. It does not capture the concept of desert. It does not work for earning either.[32]

EARNING AS REDEMPTION

An unearned opportunity is an unearned opportunity, but though unearned, a person may yet do justice to it. That possibility often is what we have in mind when we say a person deserves a chance. To ignore that possibility is to ignore the possibility of redemption involved in working to do justice to an opportunity.

In a deservedly popular film about World War II, *Saving Private Ryan*, Captain Miller is fatally injured while rescuing Private Ryan. As Miller dies, he says to Ryan: "Earn it!" At that moment, neither character is under any illusions about whether Ryan earned the rescue. He did not, as they both know. Neither is Ryan choiceworthy in the sense of element (b), as they both know. (As the story goes, the reason why High Command orders Ryan's rescue has nothing to do with Ryan's worthiness. Ryan's three brothers have just died in battle. The point of rescuing Ryan is to avoid having to send a telegram to Ryan's mother saying her entire family has just been wiped out.) Still, as both characters also know, that is not

[32] However, we might defend a version of Feldman's (1995) thesis in this way. The soldier, awarded a medal in advance, does not deserve it and has not earned it. (The medal is an award, not an opportunity. If it is deserved at all, it must be deserved qua award, which is to say it must be deserved along lines specified by the compensatory model.) Even so, it can make sense to honor the soldier now for what the soldier is about to do. Then, after the soldier makes the heroic sacrifice, it will make sense to speak of the soldier as having earned the medal.

the end of the story, for it is now up to Ryan to settle whether Miller's sacrifice was in vain.[33] It is not too late for Ryan to try to redeem the sacrifice by going on to be as worthy as a person could be.[34]

If there is anything Ryan can do to earn the rescue, it will be at t_2, not t_1, as analogous to the promissory model's element (a). When Miller says, "Earn it" he fully realizes that Ryan has not yet done his part. Ryan's rescue can never be deserved in the way a reward or prize is deserved. To be earned (deserved) at all, the rescue will have to be earned in the way advance salary is earned: that is, after the fact. Fittingly, the film ends with a scene from decades later. An elderly Ryan visits Miller's grave. Anguished, Ryan implores his wife to, "Tell me I've been a good man!" The implication: If Ryan has been a good man, then he has done all he could to earn the rescue that gave him a chance to be a good man.

DOING JUSTICE TO OUR OPPORTUNITIES

Notice that Ryan's story is neutral regarding the relevance of alternative desert bases. The elderly Ryan's wife may say the relevant basis is effort and thus that Ryan is deserving in virtue of having done all he could. Ryan himself may see achievement as the relevant basis, and conclude that despite his efforts he has not done nearly enough to be worthy of all the lives sacrificed to save his. The problem is general. If great sacrifices were made to put us in a position to flourish, we have to wonder whether there is anything we can do to be worthy of those sacrifices. The easy answer is that if we do all we can, we have done all anyone could ask. Yet, if we are reflective, we cannot help but think the easy answer sometimes is too easy, and that there is no guarantee that "doing all we can" will be enough.

Good luck cannot rob us of the chance to act in ways that make people deserving, although bad luck can, which is one reason why bad luck is bad. For example, if Private Ryan is killed by a stray bullet within minutes

[33] Abraham Lincoln's Gettysburg Address, one of the most moving speeches ever made, gains its rhetorical power from precisely this point, speaking as it does of the unfinished work of those who died in battle, calling on us to make sure their last full measure of devotion shall not be in vain.

[34] Here is another way of interpreting what Captain Miller means when he says, "Earn it." Miller is saying Ryan owes it to the men who died to be as worthy as possible of their sacrifice. So interpreted, Miller's question invokes compensatory as well as promissory models. Going on to be as worthy as possible is the closest Ryan can come to giving the fallen soldiers what they deserve in recognition of their sacrifice. I owe this thought to an e-mail exchange with Bas van der Vossen.

of being rescued, then there is no fact of the matter about whether Ryan did justice to the opportunity to live a good life, since (in this example) he got no such opportunity. Bad luck robbed him of it.

In some ways, Ryan's situation is like a lottery winner's. If Miller hands Ryan a winning lottery ticket and says with his dying breath, "Earn it," can Ryan earn it? No one would say Ryan has earned it at t_1,[35] but that is not the end of the story, because even when a windfall is sheer luck, it is not only sheer luck. It is also a challenge, and as with most challenges there is a right way of responding. Some day, there will be a fact of the matter regarding whether Ryan responded well.

Private Ryan's situation also is a bit like that of persons born with natural and positional advantages. We are not born having done anything to deserve advantages as rewards. So, a standard compensatory model has no resources that could underwrite claims of desert at the moment of birth. At birth, we are merely lucky. Also, at birth, there is no basis for deeming us choiceworthy, if choosing were even an issue. Thus, the promissory model's element (b) likewise is unable to underwrite claims of desert at the moment of birth. Still, regarding our advantages, there is something we can do later on, in the manner of element (a). We can do justice to them.

SUMMARY

The import of the promissory model's element (a) is that what once was morally arbitrary need not remain so. The most valuable things we are given in life are opportunities, and the main thing we do to deserve them is to do justice to them after the fact. The import of element (b) is that this theory has room for the commonsense idea that people can deserve a chance. They can deserve a chance not because of what they have done but because of what they can and will do, if only we give them a chance.

It is crucial that the scales be balanced. It is not crucial that components of the balance be supplied in a particular order. If X is conferred first, and the desert base is supplied later, that too is a balancing of the moral scale.

[35] If the case were more like the kind of case covered by element (b), Captain Miller conceivably might say Ryan deserves the ticket. For example, suppose Miller needs to select someone from a list of applicants, and sees that Ryan would move mountains to prove himself worthy. In that case, deeming Ryan choiceworthy on that basis might be Miller's best-justified option.

DISCUSSION

Thomas Nagel says "It would be difficult to dispute Rawls's claim that no one deserves his greater natural capacity nor merits a more favorable starting place in society, except perhaps by appealing to a theory of the transmigration of souls."[36] Is this as obvious as Nagel thinks? Or, should we think that, because people can do justice to their opportunities, what initially was undeserved need not remain so?

[36] Nagel 1997, 309.

10

Grounding Desert

THESIS: One justification for giving people credit for using their opportunities well is that doing so empowers people to use their opportunities well, thereby helping them to live well together.

WHY ONE CONCEPTION RATHER THAN ANOTHER?

Are the two models, compensatory and promissory, truly models of desert? Does it matter? The main issue is not whether we use the same word when referring to those who did their best before receiving rewards and to those who did their best after receiving opportunities. In fact, we do, but the larger question is, are we *justified* in thinking of desert claims as weighty in both cases?

I explained how in everyday life we grasp the concept of deserving a chance in virtue of what we did, or will do, with it. I would not appeal to common sense to justify our common-sense understanding, though. To justify, we look elsewhere. This chapter indicates (although only indicates) where we might look.

Part of what makes it difficult even to begin such a discussion is that, in trying to justify, we risk trivializing. We risk seeming to ground a thing in considerations less important than the thing itself. That could be a problem when trying to justify a conception of justice. When assessing alternative conceptions of justice, we generally cannot settle the contest by appeal to yet another lofty but contested ideal of justice. However, if we appeal to something else – something other than (our conception of) justice – we are bound to be appealing to that which seems less important. But that is okay. We are not confusedly seeking the foundation of that which is itself foundational. We simply ask what can be said on the conception's behalf.

THE LEAST ADVANTAGED

What they want to get; what they want to be

Margaret Holmgren says justice "demands that each individual be secured the most fundamental benefits in life compatible with like benefits for all," then adds, "The opportunity to progress by our own efforts is a fundamental interest."[37] Richard Miller concurs: "Most people (including most of the worst off) want to use what resources they have actively, to get ahead on their own steam, and this reflects a proper valuing of human capacities."[38]

On one view, the Rawlsian supposition that inequalities should be arranged to benefit maximally the least advantaged rules out the idea that people deserve more – and thus should get more – if and when and because their talents and efforts contribute more to society. Holmgren, though, notes that people in Rawls's original position would know (because by hypothesis they are aware of perfectly general features of human psychology) that people not only want to be given stuff; they want to be successful, and they want their success to be deserved. Accordingly, even grossly risk-averse contractors, focusing only on the least advantaged economic class, would be nonetheless anxious to ensure that such people have opportunities to advance by their own effort. "Rather than focusing exclusively on the share of income or wealth they would receive, they would choose a principle of distribution which would ensure that they would each have this opportunity."[39]

Holmgren's claim seems incompatible with Rawls's difference principle *if* we interpret the principle as Nozick interprets it, as a ground-level prescription for redistribution. In that case, the idea that Jane deserves her salary threatens to override our mandate to lay claim to her salary on behalf of the least advantaged. However, Nozick's way is not the only way to interpret the difference principle. Suppose we interpret the principle not as a mandate for redistribution but rather as a way of evaluating basic structure. That is, we evaluate basic structure by asking whether it works to the benefit of the least advantaged. On the latter interpretation, we

[37] Holmgren 1986, 274.
[38] Richard W. Miller 2002, 286.
[39] Holmgren 1986, 275.

choose among rules such as "try to give people what they deserve" and "try to give the least advantaged everything" by asking which is best for the least advantaged in actual empirical practice.

The latter undoubtedly is the difference principle's canonical interpretation. Unfortunately, we naturally slip into thinking of bargainers as choosing a plan for redistribution. Rawls himself slips in this way when he says, "There is a tendency for common sense to suppose that income and wealth, and the good things in life generally, should be distributed according to desert.... Now justice as fairness rejects this conception. Such a principle would not be chosen in the original position."[40] We can agree that such a principle would not be chosen, but the reason is because distributional principles per se are not on the menu. They are not even the *kind* of thing bargainers choose. Bargainers choose meta-level principles for *evaluating* principles like distribution according to desert.

Read in this canonical way, the difference principle, far from competing with principles of desert, can *support* the idea that people deserve a chance. The difference principle supports principles of desert if Holmgren is correct to say the least advantaged want and need the chance to prosper by their own merit. Likewise, the difference principle supports principles of desert if it is historically true that the least advantaged tend to flourish within, and only within, systems in which honest hard work is respected and rewarded. Such a system may be the best that unskilled laborers could hope for: best for them as wage laborers, as consumers of what other workers produce, as parents who believe their children deserve a chance, and perhaps also as people who may one day need the kind of safety net (private or public insurance) that only a healthy economy can afford. Rawlsians and non-Rawlsians alike can see these considerations as weighty.

Is the Original Position an Experience Machine?

Robert Nozick's "Experience Machine" lets us plug our brains into a computer programmed to make us think we are living the best life possible. The life we think we are living is a computer-induced dream, but we do not know that. Whatever experience would be part of our felt experience. Nozick asks, "Would you plug in? *What else can matter to us, other than how our lives feel from the inside?*"[41]

[40] Rawls 1971, 310.
[41] Nozick 1974, 43 (emphasis in original).

Nozick does not connect the question to his critique of Rawls, yet we may at least wonder whether the less advantaged want to accomplish things, not simply be given things with which they can go on to have pleasant experiences. That is, we can ask of the less advantaged, how many of their children grow up to register patents? Earn PhDs? Earn a living as professional musicians? Does Rawls's difference principle ask these questions, or is it concerned only that the less advantaged have a pleasant experience? (see chapter 31)

HOW TO CARE ABOUT CONSEQUENCES

Likewise, utilitarians and nonutilitarians alike can care about consequences. Feinberg says, "The awarding of prizes directly promotes cultivation of the skills which constitute bases of competition."[42] Rawls says, "Other things equal, one conception of justice is preferable to another when its broader consequences are more desirable."[43] Although these thinkers are not utilitarians, it does not stop them from making factual claims about the consequences of respecting desert claims (in Feinberg's case), or from noting that good consequences are better than bad ones (in Rawls's case).

Both Feinberg and Rawls can correctly insist that utility is not a desert maker, while also recognizing that (a) things that are desert makers (effort, excellence) can as a matter of fact make people better off, and that (b) making people better off is morally significant. Rachels adds, "In a system that respects deserts, someone who treats others well may expect to be treated well in return, while someone who treats others badly cannot. If this aspect of moral life were eliminated, morality would have no reward and immorality would have no bad consequences, so there would be less reason for one to be concerned with it."[44]

In short, our ordinary notions of desert serve a purpose. One (if only one) way in which a society benefits people is by distributing fruits of cooperation in proportion to contributions to the cooperative effort. That is how societies induce contributions to begin with. Desert as normally understood is part of the glue that holds society together as a productive venture. Respecting desert as normally understood (respecting the inputs people supply) makes people in general better off. To be sure, it

[42] Feinberg 1970, 80.
[43] Rawls 1971, 6.
[44] Rachels 1997, 190.

would be a misuse of terms to say Bob deserves a pay raise on the grounds that giving him a raise would have utility. We may say Bob deserves a raise because he does great work, does more than his share, and does it without complaint. We do not say giving Bob a raise would have utility. But if we ask why we should *acknowledge* that Bob is a great worker, a big part of what makes Bob's efforts worthy of recognition is that his efforts are of a kind that make us all better off. If we ask why Bob is deserving, the answer should be: Bob supplied the requisite desert makers. If we ask why we *care* whether Bob supplied inputs that make a person deserving, one answer would be: Supplying those inputs makes Bob the kind of person we want our neighbors, our children, and ourselves to be, and makes us all better off to boot.

The point need not be to maximize utility so much as to show respect for customs and institutions and characters that make people better off. (Either way, desert tracks *constructive* effort rather than effort per se. Effort tokens need not be successful, but they do need to be of a type that tends to produce worthy results.) If we are to do justice to individual persons, then when their individuality manifests itself in constructive effort, we had better be prepared to honor that effort, and to respect the hopes and dreams that fuel it.

DESERVING AND NEEDING

When we say "she deserves a chance," how does that differ from saying she *needs* a chance? "Deserves" suggests she has some realized or potential merit in virtue of which she ought to be given a chance, whereas "needs" suggests neither real nor potential merit. However, when we say, "*all* she needs is a chance," that comes close to saying she deserves a chance. It comes close to saying she is the kind of person who will give the opportunity its due.

Nonetheless, whatever room we make for desert, the fact remains that people's needs matter, at least at some level.[45] I would go so far as to say desert matters partly because needs matter. That Bob needs X is no reason to say Bob deserves X for the same reason that X's utility is no reason to say Bob deserves X. And if that is true, then need is not a desert base. But there are other ways for need to be relevant.

Suppose for simplicity's sake that the only way to deserve X is to work hard for X. In that case, by hypothesis, need is not at all relevant to

[45] I am agreeing here with, among others, Brock 1999, 166.

whether Bob deserves X. By hypothesis, all that matters is that Bob worked hard for X. Still, even though by hypothesis need has nothing to do with our reason for thinking Bob deserves X, need remains a reason for *caring* about desert. One reason to give people what they deserve is that it renders people willing and able to act in ways that help them (and the people around them) to get what they need. Welfare considerations are not desert bases, but they can still provide reasons for taking a given desert maker seriously (for example, for respecting people who work hard).

A KANTIAN GROUNDING

When wondering whether a person did justice to an opportunity, we typically do not look back to events occurring before the opportunity was received. I indicated how we might argue for this on consequentialist grounds. It may be a good thing on Kantian grounds too. Although I will not press the point, there is something necessarily and laudably ahistorical about simply respecting what people bring to the table. We respect their work, period. We admire their character, period. We do not argue (or worse, stipulate as dogma) that people are products of nature/nurture and thus ineligible for moral credit. Sometimes, we simply give people credit for what they achieve, and for what they are. And sometimes, simply giving people credit is the essence of treating them as persons rather than as mere confluences of historical forces.

Part of the oddity in doubting whether Jane deserves her character is that Jane's character is not something that happened to her. It *is* her. Or if we were to imagine treating Jane and her character as separate things, then it would have to be Jane's character that we credit for being of good character, so the question of why Jane per se should get the credit would be moot. In truth, of course, it is people, not their characters, that work hard. Thus, if we say exemplary character is morally arbitrary, it is people, not merely character, that we are refusing to take seriously.

Martin Luther King once said, "I have a dream that my four children will one day live in a nation where they will not be judged by the color of their skin but by the content of their character."[46] This was a dream worth living and dying for. King did not dream his children would live in a nation where their characters would be seen as accidents for which they could claim no credit. King asks us to judge his children by the content of

[46] In his "I Have a Dream" speech.

their character, not by its causes. That was the right thing to ask, because that is how we take characters (that is, persons) seriously.

If the characters of King's children are not taken seriously, they will get neither the rewards nor the opportunities they deserve. Especially by the lights of Rawls's difference principle, this should matter, for the least advantaged can least afford the self-stifling cynicism that goes with believing no one deserves anything. Neither can they afford the license for repression that goes with the *more* advantaged believing no one deserves anything.

These remarks indicate that the possibility of deserving a chance is not mere common sense. In the end, the bottom line is in part a practical question, somewhat amenable to empirical testing: Which way of talking – about what people can do to be deserving – empowers people to make use of their opportunities?

11

Desert as Institutional Artifact

THESIS: Desert has institutional and preinstitutional aspects.

JUSTICE – INSTITUTIONAL AND NATURAL

To Feinberg, "desert is a natural moral notion (that is, one which is not logically tied to institutions, practices, and rules)."[47] Rawls denies that desert is natural in this sense, but concedes the legitimacy of desert claims as institutional artifacts. Thus, faster runners deserve medals according to rules created for the express purpose of giving medals to faster runners. Those who "have done what the system announces it will reward are entitled to have their expectations met. In this sense the more fortunate have title to their better situation; their claims are legitimate expectations established by social institutions and the community is obligated to fulfill them. But this sense of desert is that of entitlement. It presupposes the existence of an ongoing cooperative scheme...."[48] The idea is that at

47 Feinberg 1970, 56.

48 Rawls 1999a, 89. In the 1971 edition, the final sentence reads, "But this sense of desert presupposes the existence of an ongoing cooperative scheme" (1971, 103). So the explicit assimilation of desert to entitlement came later. However, the next paragraph of the 1999 edition makes a further change that goes in the opposite direction, as if unaware of the change to the previous paragraph. Rawls says in that next paragraph that we do not deserve our social endowments, or even our character, "for such character depends in good part upon fortunate family and social circumstances in early life for which we can claim no credit. The notion of desert does not apply here. To be sure, the more advantaged have a right to their natural assets, as does everyone else" (1999a, 89). The last sentence is a new addition, separating desert, which does not apply, from entitlement, which does.

some point we will be in a position to define, then acknowledge, claims of desert, but such claims (1) will have no standing outside the context of particular institutional rules, and therefore (2) cannot bear on what rules we should have in the first place.[49]

Other senses of desert, though, are less closely tied to institutional structures. A medalist who trains for years deserves admiration in a way that a medalist who wins purely on the strength of genetic gifts does not, even when the two are equally deserving of medals by the lights of the institutional rules. Likewise, athletes prove themselves worthy of the faith of their families and coaches by doing all they can to win, and by being role models in the process, even when institutional rules are silent on the relevance of such inputs.

If we actually believed existing institutional rules are all that matter so far as desert is concerned, we would need to explain why we so readily criticize rules of particular institutions, and outcomes that accord with extant rules. Canadian sprinter Ben Johnson ran the fastest time in the hundred meter race at the 1988 Olympics. He did nothing to show that he deserved his genetic gifts, or his competitive character, or the excellence of his coaches. All he did was run faster than the competition, which on its face entails he deserved the gold medal.

However, blood tests revealed that Johnson had taken steroids. Did it matter? Yes it did. The fact that he took steroids raised questions of desert, whereas the bare fact that Johnson had a *background* (he had genes; he grew up in an environment) did not. Being born in the wake of the Big Bang did not stop Johnson from deserving a medal, but there is a real question about whether taking steroids preempts inputs by which sprinters come to deserve medals. We may ask whether steroids are in fact banned. That is an institutional question. We also may ask whether steroids *should* be banned. That question is preinstitutional: its answer (1) does not turn on particular institutional rules, and (2) does bear on what rules we should have in the first place.

THE PREINSTITUTIONAL ASPECT

As noted, Rawls says those who do what the system announces it will reward are entitled to have their expectations met. Rawls insists the status of such expectations is an institutional artifact. He is right in one way and wrong in another. On one hand, it is an institutional artifact that the winner is

[49] Rawls 1971, 103.

entitled to a gold rather than a platinum medal. On the other hand, it is a preinstitutional moral fact that *if* the system promises a gold medal to the winner, then the system ought to give the winner a gold medal.

Notice: The system need not *announce* an obligation to keep promises. It has that obligation regardless. Therefore, while many of the factors that go into determining entitlements may be institutional artifacts, this one is not.

Obviously, some desert claims carry moral weight as institutional artifacts. (It makes sense for a winner to claim to deserve a platinum medal only if that is what the system has led the winner to expect.) However, some claims do not merely *happen* to carry weight as institutional artifacts. They *should* carry weight as institutional artifacts because they carry weight preinstitutionally. It is a matter of indifference whether the system promises the winner a gold medal or a platinum medal. It is not a matter of indifference whether the system encourages excellence rather than corruption or incompetence. We see winning sprinters as deserving when we see their excellence as a product of years of ferocious dedication. If instead we thought the key to winning was more drugs, we would not regard winners as deserving. This difference is not an institutional artifact. We see the cases differently even when a performance-enhancing drug is allowed by the rules.

Part of our reason for caring is that the race's point is to show us how excellent a human being can be. If we explain success in terms of steroids rather than in terms of features of persons that ground desert claims in a preinstitutional sense, the institution is not working. If the competition inspires impressionable viewers to take steroids rather than to develop their talents, the institution is not working. If one way of competing risks competitors' lives and sets a dangerous example for children who idolize them, while a version that bans steroids is healthier for everyone, then we have preinstitutional grounds for thinking it was right to establish, publicize, and enforce the ban, and that my countryman Ben Johnson did not deserve a medal.[50]

PUZZLES

1. How often do we find ourselves unsure which model (compensatory or promissory) is more relevant? For example, how do we

[50] This conclusion does not presuppose the promissory model. The possibility of preinstitutional desert is manifest even within the compensatory framework.

know whether to see an award as a reward or as a challenge to live up to? Suppose Jane gets a scholarship, then decides to treat it as a challenge, something she must strive to be worthy of. Could Jane be wrong? What would make her wrong? If the awarding agency explicitly stipulates that Jane's award is in recognition of her past performance, does that settle the matter? What if Jane chooses to see the award differently? Must Jane be in error?

2. David Miller says, suppose I "persuade the local archery club to let me take part in its annual tournament. By sheer good luck I send three arrows into the gold, something I could not repeat in a million attempts. I could not on *this* basis deserve the trophy that is presented to me."[51] Is Miller right? By the institutional rules, all that matters is where the arrows land, so does Miller's example prove that ordinary thinking about desert has a preinstitutional component?

[51] David Miller 1999a, 134. Emphasis added.

12

The Limits of Desert

THESIS: Conceptions of desert respond to people as *active* agents. Conceptions of entitlement respond to people as *separate* agents.

WHOSE PERFORMANCE IS IT?

Did Wilt Chamberlain do justice to the potential given to him by luck of the draw in the natural lottery?[52] One possible answer is that whether he did justice to his potential is no one else's business. Wilt is not *indebted* to anyone for his natural assets. He did not borrow his talent from a common pool. No account is out of balance merely by virtue of Wilt having characteristics that make him Wilt. Still, even if it is no one else's business whether Wilt does justice to his potential, the fact remains that one way or another, Wilt will do, or fail to do, justice to it.

Part of our reason for thinking it is Wilt rather than you or me who deserves credit for the excellence of Wilt's performance is that, as David Miller puts it, "the performance is entirely his."[53] Note: The issue is not whether the performance is Wilt's rather than the Big Bang's; the issue is whether the performance is Wilt's rather than some other person's. The question of whether to credit Wilt for his performance is never a question of whether Wilt caused himself to have his character and talent. Instead, the question is whether the character, talent, or

[52] I thank Paul Dotson and Peter Dietsch for discussions about what is involved in having status as a person.

[53] David Miller 1999a, 144.

other desert-making inputs are, after all, Wilt's, rather than some other person's.[54]

If and when we applaud Wilt's effort, we imply that the credit is due to Wilt rather than to, for example, me. Why? Not because Wilt deserved the effort (whatever that would mean) but because the effort was Wilt's rather than mine. When we ask whether the effort is truly Wilt's, the answer sometimes is simply yes. Other times, we may want to credit Wilt's coaches, teachers, or parents, for performances that contributed to Wilt's in tangible ways.[55]

Notice that giving credit is not a zero-sum game. We do not think less of Wilt when Wilt thanks his parents. Indeed, we think less of Wilt if he fails to give credit where credit is due. The credit due to Wilt's parents takes away from credit due to Wilt only if the implication is that the performance we thought was Wilt's was not really his. (Imagine Wilt, in an acceptance speech for an academic award, thanking his parents and coaches for writing all those term papers.)

HOW TO SEE DESERT AS NONCOMPARATIVE

Desert is not essentially a comparative notion. In particular, the models of desert developed here make room *not* for honoring those with advantages as compared to those without, but for honoring people who do what they can to be deserving of their advantages. These elements of a larger theory of justice ask whether a person has supplied the requisite desert makers,

[54] Charles Beitz says, "While the distribution of natural talents is arbitrary in the sense that one cannot deserve to be born with the capacity, say, to play like Rubinstein, it does not obviously follow that the possession of such a talent needs any justification. On the contrary, simply having a talent seems to furnish prima facie warrant for making use of it in ways that are, for the possessor, possible and desirable. A person need not justify the possession of talents, despite the fact that one cannot be said to deserve them, because they are already one's own: the prima facie right to use and control talents is fixed by natural fact" (1979, 138).

[55] In this way, when we get to the bottom of desert, it turns out to presuppose a rudimentary conception of entitlement, or at least possession. We must have a sense of when a talent is mine and not Wilt's.

It may go both ways. For example, when Locke speaks of how an unowned thing becomes property when we mix labor with it, the idea seems to be that if a thing is unowned, then to connect it to us via desert-making effort results in a relationship that others cannot ignore without ignoring what we deserve. That would not suffice to confer title to a thing already owned by someone else, but it does put us in a position where someone who takes a previously unowned thing without our consent would be wronging us. Thus, on this Lockean view, when we get to the bottom of entitlement, it turns out to presuppose a rudimentary conception of desert.

not whether the person has done more than someone else has. There are cases like the following:

a. Wilt Chamberlain has X and you have Y,
b. Wilt did something to deserve X while you did something to deserve Y,
c. X is more than Y, and yet (so far as desert is concerned),
d. there is nothing wrong with X being more than Y, despite the fact that Wilt does not deserve "more than you" *under that description.*

In other words, the question about Wilt is not whether Wilt did something to deserve *more than you,* but whether Wilt did something to deserve *what he has.* Perhaps there was never a time when an impartial judge, weighing your performance against Wilt's, had reason to conclude that Wilt's prize should be larger than yours. All that happened is that Wilt did justice to his opportunity and you did justice to yours. Should we focus on the relation, or imagine there is one, between you and Wilt, or should we focus on a pair of relations, one between what Wilt did and what Wilt has, and a second between what you did and what you have? Perhaps neither focus captures the whole truth about justice, but the second focus (that is, on the pair of relations) is a focus on desert, where the first is a focus on something else, something more comparative, such as equality.[56]

A central distributor, intending to distribute according to desert, would need to judge relative deserts, then distribute accordingly. If there is no central distributor, the situation is different. If Wilt worked hard for his salary of X while you worked hard for your salary of Y, there is something fitting in Wilt getting X and you getting Y. You each supplied desert-making inputs connecting you to your respective salaries. It might be hard for a central distributor to justify judging that Wilt deserves so much more than you, but by hypothesis there was no such judgment.

HOW MUCH SHOULD WILT BE PAID?

Needless to say, Wilt deserves no credit for the economic system that attaches a given salary to Wilt's performance. On the other hand, Wilt does not need to deserve credit for the system. He claims credit only for his performance. If it is Wilt rather than you who deserves credit for Wilt's

[56] Olsaretti (2004, 166–8) argues that theories of desert cannot easily justify inequality. She is right, not because theories of desert *fail* in their attempt to justify inequality but because they make no such attempt. They do not presume inequality needs justifying.

performance, then it is Wilt rather than you who has a presumptive claim to the salary that the system (or rather, Wilt's employer) attaches to Wilt's performance.

You may doubt Wilt's profession should be paid so much more than yours, not because you think people in top professions are undeserving, but because you think there is a presumption against that much inequality. You may think no amount of desert could be enough to overturn that presumption. You may be right. It would have to be argued within the context of a theory of equality, which reminds us that we need to keep our conclusions regarding desert in perspective. What I call "deserving a chance" is not the whole of desert. Desert is not the whole of justice. Justice is not the whole of morality. This part of a larger theory tells us to treat opportunities as challenges and to respect those who meet their own challenges in fitting ways, but this part *does not answer all questions.* It does not say what Wilt Chamberlain should have been paid, or what opportunities Wilt should have had. It answers *one* question: What can Wilt or anyone blessed by good fortune do to be deserving? Its answer is: When we look back on Wilt's career, wondering whether he deserved his advantages, we are not restricted to considering what he did before receiving them. What matters, if anything at all matters, is what he did with them.

DESERT VERSUS ENTITLEMENT

Our reasons to respect desert as normally understood also are reasons to respect desert's limits as normally understood. In particular, there are limits to what a society can do, and limits to what society can expect its citizens to do, to ensure that people get what they deserve. Thus, even something as fundamental as the principle that people should get what they deserve has limits.

A just system works to minimize the extent to which people's entitlements fly in the face of what they deserve, but not at a cost of compromising people's ability to form stable expectations regarding their entitlements, and thus to get on with their lives in peaceful and productive ways. The point goes both ways, though, for desert in turn corrects the caprices of rightful entitlements, and that too is a good thing. For example, a proprietor may know her employee is entitled to a certain wage while also seeing that the employee is exceptionally productive and (in both promissory and compensatory senses) deserves a raise. If she cares enough about desert, she restructures her holdings (her payroll) accordingly, benefiting not only the employee but probably her company and her customers as well.

Principles of entitlement acknowledge our status as *separate* agents. Principles of desert acknowledge our status as *active* agents. A society cannot work without a "rule of law" system that secures people's savings and earnings, thereby enabling people to plan their lives.[57] Neither can a rule of law function properly in the absence of an ethos that deeply respects what people can do to be deserving.[58] Part of our job as moral agents is to do justice to opportunities embedded in our entitlements. It is in meeting that challenge that we make entitlement systems work.

57 Waldron 1989.

58 What determines whether a given salary is a fitting response to the desert-making inputs we supply? In the abstract, a theory of desert cannot say. Salaries are artifacts of systems of entitlement, and systems of entitlement are not pure responses to facts about what workers deserve. They also respond to notions of reciprocity, equality, and need, and to all kinds of factors (supply and demand) not directly related to matters of justice. Thus, the going rate for a type of work will not be determined by what a particular worker deserves, although whether a worker deserves to be paid the going rate will depend on whether that worker does something (supplies the expected desert-making inputs) to deserve it.

Some notions of desert are defensible by virtue of encouraging us to respect mutually advantageous systems of entitlement (see Chapter 24 and, generally, Part 5). Some notions of entitlement are defensible by virtue of empowering us to do something to deserve such opportunities as come our way. Which notion is more foundational? Out of context, there is no truth of the matter. In the context of aiming to justify a notion of desert, we must treat something else as foundational, if only for argument's sake. Likewise with entitlement. What we aim to justify defines the context and determines what can and what cannot be treated as foundational.

PART 3

HOW TO RECIPROCATE

13

Reciprocity

> The prosecutor was astounded. Under her breath, she seethed. "Six months probation! The guy's as guilty as sin! He should do five years hard labor!"
>
> The judge spoke again to a stunned courtroom. "This is a court of justice, and I have above all sworn to uphold principles of justice. In passing sentence, I've been guided by that solemn oath. After due deliberation I conclude that the overriding principle in this case is the principle of reciprocity. The crime of which the defendant has been found guilty is extremely serious, but the overriding consideration in determining sentence was that, well, I owed the guy a favor. I am bound by justice to pass sentence as I have. This court is adjourned."

As with desert, reciprocity inspires skepticism. On one hand, it seems obvious that when we return a favor to someone who has been good to us, we do something that (other things equal) is at least good, maybe even morally required. Yet, Allen Buchanan recently began an article by declaring, "There is a strain of thought in the history of ethics that surfaces from time to time in the work of powerful thinkers and that threatens to shatter the basic conceptual framework within which our legal system and commonsense morality formulate the problems of justice. This idea may be called justice as reciprocity."[1] To Buchanan, justice as reciprocity implies that duties of justice obtain only among those who can do each other favors.[2] If Buchanan is right, then what he calls justice as reciprocity is at best only a part of justice – a part that is silent on duties between people who have no favors to offer each other.

[1] Buchanan 1990, 227. Buchanan's main foil here is David Gauthier.

[2] Buchanan 1990, 228.

Still, the more modest root idea of reciprocity – the idea that returning favors is at very least a good thing – remains compelling. What can we say on behalf of this root idea? Chapter 14 defines reciprocity, particularly by contrast to principles of desert. Chapter 15 explores variations on the theme of reciprocity that do not threaten to shatter commonsense morality but are instead indispensable parts of it. Chapter 16 considers whether obligations to society can be grounded in reciprocity. Chapter 17 asks when such obligations are enforceable, while exploring more generally the question of reciprocity's moral limits.

14

What Is Reciprocity?

Thesis: Principles of reciprocity can play important roles in a pluralistic theory of justice.

RECIPROCITY, DESERT, SELF-RESPECT

Lawrence Becker, in a wonderful, neglected book, calls reciprocity a disposition "to return good in proportion to the good we receive, and to make reparation for the harm we have done. Moreover, reciprocity is a fundamental virtue. Its requirements have presumptive authority over many competing considerations."[3] The disposition is ubiquitous. "Gifts and goods pervade our lives. So do evils and injuries. Everywhere, in every society of record, there is a norm of reciprocity about such things."[4]

> The details differ strikingly from place to place, time to time, and every society is profuse with forms. There are rituals of gift-giving, unspoken understandings between lovers, patterns of family life, expectations among friends, duties of fair play, obligations of citizenship, contracts – all understood as reciprocal. There is an intricate etiquette for it all, and it is connected (both in theory and in practice) to prudence, self-interest, altruism, basic human needs, social welfare, notions of desert and duty, justice, and fairness.[5]

[3] Becker 1986, 3. Becker's book is so rich that it is hard to say anything useful about the topic that is not to some degree anticipated there.

[4] Becker 1986, 73.

[5] Becker 1986, 73.

Reciprocity concerns how we should respond when someone has done us a favor. To Becker, reciprocity so understood is "fundamental to the very concept of justice."[6] Formulated as a principle, the idea might be:

When you can, return good in proportion to the good you receive.

Sometimes, we can be precise about what qualifies as a proportionate return.[7] If someone lends you twenty dollars, that calls for you to repay the loan, and to be disposed to offer a similar favor in similar circumstances. However, if you take out a second mortgage to rescue someone from bankruptcy, you will not be doing it as a way of returning the favor of a twenty dollar loan. Thus, there are real, albeit vague, boundaries; some gestures would be too little, others too much.

At times, we ask not only whether a gesture is of an appropriate magnitude, but also whether it is of an appropriate kind. If your friend Jones drives you to the airport, and you try to reciprocate by giving Jones twenty dollars, there may be nothing wrong with the gesture's magnitude. Twenty dollars may be about right in terms of magnitude; there is no number of dollars that would be better. The problem is, dollars typically are the wrong *kind* of response to favors done by a friend. Successfully returning a favor involves responding to the spirit as well as to the magnitude.

Note that wanting to do something in return does not mean being obsessed with getting out of debt. The art of reciprocity is partly an art of graciously acknowledging favors. Sometimes, we simply give thanks, without meaning to imply that we are settling the account. Another part of the art has to do with timing, since wanting to do something need not mean wanting to do something immediately. Among friends, we reciprocate not on a per transaction basis but over the long run, with a view to the pattern of the whole relationship. (For better or worse, this fact is central to our most intimate relations. We are creatures of habit. In intimate relationships, unless we take stock from time to time, our habits tend to devolve into ways of exploiting or being exploited by, and ultimately alienating, our partner.)

[6] Becker 1980b, 417.

[7] To complicate matters, there can be more than one salient proportion. Do we want to match the benefit we received, or the burden our benefactor bore in doing us a favor? At times, we cannot do the one without going overboard on the other. I thank Chris Brown for the observation.

A principle of reciprocity features prominently in James Rachels's illuminating discussion of what people deserve. Rachels asks us to think about the following case.

The ride to work. You, Smith, and Jones all work at the same place. One morning your car won't start and you need a ride, so you call Smith and ask him to come by for you. But Smith refuses. He doesn't want to be bothered, so he makes up some excuse. Then you call Jones, and he gives you the ride you need. A few weeks later, you get a call from Smith. Now he is having car trouble and he asks you for a ride. Should you help him?[8]

Should you? Two conclusions suggest themselves. First, as Rachels concludes, Smith does not deserve to be helped, so far as we can tell. Second, you do not owe Smith a ride as a matter of reciprocity. Rachels does not draw the second conclusion, but presumably would endorse it.

We can go farther. Suppose we change Rachels's case so that, a few weeks later, it is Jones rather than Smith who calls for help. That is a different situation, in at least two ways. First, Jones arguably deserves help because, if your experience with him is any indication, Jones is the kind of person who helps people. Notably, this reason to help counts as a reason for anyone. Second, because Jones helped *you*, you in particular have an additional reason to help, namely that your coming to Jones's aid would repay a favor.[9] This second reason responds not to the kind of person Jones is, but to the history you and Jones share. You in particular are in his debt.

Rachels himself goes on to modify the case in a different way.

Simultaneous requests. Smith calls and asks for a ride. Meanwhile, Jones is on the other line also needing a ride. But they live in opposite directions, so it is impossible for you to help both. Which do you help?[10]

Rachels finds it obvious you should help Jones. Why? Because, Rachels says, Jones is more deserving. We can agree; you should help Jones. I wish only to add: Something else is going on here. Yes, Jones seems more deserving, but in addition, Smith and Jones are calling *you*, and you in particular have a history with each.

Suppose Smith and Jones instead call Bloggs. Bloggs has never done a favor for either, nor asked for one, and thus the only issue for Bloggs (we may suppose) is that Jones is more deserving. If Bloggs helps Smith

[8] Rachels 1997, 189.
[9] Rachels 1997, 189.
[10] Rachels 1997, 190.

and leaves Jones stranded, then Smith gets more than he deserves while Jones gets less. As Rachels actually tells the story of *Simultaneous Requests,* though, Smith and Jones call you, not Bloggs. If it is you, not Bloggs, who helps Smith instead of Jones, then there is a further issue. If you help Smith instead of Jones, you ignore not only what they deserve but also what you *owe* to Jones. Your failure is more than a failure to give Jones what he deserves. It is a failure to honor a debt. As a matter of self-respect, it should have mattered to you that Jones treated you well when Smith chose not to.

Note in passing that self-respect and raw self-interest need not coincide. If I can further my career by currying Smith's favor, while having little to gain by currying Jones's, I am still failing to respect myself (as the saying goes, I am "swallowing my pride") if, for career's sake, I ignore the history that Jones and I share.

All of this is compatible with Rachels's *Principle of Desert*: "People deserve to be treated in the same way that they have (voluntarily) treated others. Those who have treated others well deserve to be treated well in return, while those who have treated others badly deserve to be treated badly in return."[11] While this principle arguably is fine as far as it goes, it notably does not distinguish between you and Bloggs. The principle does not capture the debt to Jones that you have, but Bloggs does not, in virtue of the fact that it was you that Jones chose to help in the past.

We could modify Rachels's principle to take such debts of reciprocity into account, but we need not. Rachels's principle is no more than what it claims to be: a principle of desert. The special debt to Jones that you have, but Bloggs does not, is better captured under the rubric of a separate principle of reciprocity.

Before we move on, a word on behalf of much-maligned Smith: In the original *Ride to Work*, it was not the case that you helped Smith and Smith then refused to reciprocate. Quite possibly, if Smith had called you first and you had helped him, Smith would gladly have returned the favor. Smith's refusing to help may have been wrong, but it was not a case of refusing to return a favor. (Failing to make the first move is failing to trust, or failing to recognize an opportunity to celebrate our common humanity. Failing to make the second move is what counts as failing to reciprocate.) At least until Jones calls with a prior claim on your time, the thing to do might be to ask Smith to explain why he is calling so soon

[11] Rachels 1997, 190.

after refusing to help you. Depending on his answer, it might be right to give him the benefit of the doubt.

INDUCING COOPERATION

Interestingly, when Rachels argues that giving people what they deserve is important, he explicitly speaks of reciprocity and argues as much for reciprocity as for anything covered by his Principle of Desert. Here is what he says.

> If reciprocity could not be expected, the morality of treating others well would come to occupy a less important place in people's lives. In a system that respects deserts, someone who treats others well may expect to be treated well in return, while someone who treats others badly cannot. If this aspect of moral life were eliminated, morality would have no reward and immorality would have no bad consequences, so there would be less reason for one to be concerned with it.[12]

Rachels's argument for reciprocity's justness appeals to reciprocity's day-to-day impact on human society. The argument: Suppose we want society (or any relationship) to become and remain a cooperative venture for mutual advantage. And suppose we expect justice to help rather than hinder society in attaining this ideal. In that case, the question is not whether a conception *endorses* cooperation but whether acting on it *induces* cooperation. Returning good in proportion to good received passes this test. When people reciprocate, they teach people around them to cooperate. In the process, they not only respect justice, but foster it. Specifically, they foster a form of justice that enables people to live together in mutually respectful peace.[13]

When we consider reciprocity's pedagogical function, helping Smith and failing to help Jones are separate ways of failing to reinforce good behavior. You send each a message, and in each case, the wrong message. In terms of the risk of sending wrong messages, reciprocating benefits

[12] Rachels 1997, 190.

[13] Frances Kamm (at a workshop honoring the memory of James Rachels, Birmingham, 2004) asked me what would count as reciprocity in a case where I owe Jones a favor but meanwhile Smith and others are treating me badly. I said my owing Jones a favor is a reason to do a favor for Jones that is unaffected by how others are treating me. I may wish Smith and others would treat me better, and may wish I could do something about it, but there is no guarantee that there will be anything I can do. The world is not guaranteed to be fair in that way, and no theory should pretend otherwise. All I reasonably can aim to *guarantee* is that I *deserve* better treatment. And regarding Jones, there is something I can do qua reciprocator: I can do him the favor I owe him.

and harms seem on a par. One extra problem with reciprocating harms, though: People tend to be incompetent as receivers of negative messages. People almost never think they started the fight. A message sent as retaliation is seen by the receiver as firing the first shot, thus itself calling for retaliation, leading to cycles of violence. Like Becker, I accept that dispositions to return good for good, and bad for bad, are different dispositions. Although nothing is guaranteed in particular cases, returning good for good is something we generally can celebrate, whereas returning bad for bad is a dangerous and complex practice that no society simply can celebrate. My theory takes no position on returning bad for bad. Some of what I say might help someone to justify aspects of returning bad for bad, but I make no claim either way.[14]

PLURALISM

I defined the principle of reciprocity as saying roughly: When you can, return good in proportion to good received. I did not try to formulate the principle so as to make it immune to counterexample. The first task was to get the idea on the table. It is easy to imagine cases where reciprocity so defined would be unjust. If Jones and Smith are candidates for a job in my department, I have no right to vote for Jones as a way of returning a favor. That Jones, unlike Smith, treats people well may bear on who is the better candidate, but the fact that I owe Jones a favor does not. It is easy to understand wanting to repay a favor. What would be unjust is repaying favors with favors that are not mine to give.

So, something else is out there, in the realm of justice, limiting reciprocity's scope. There are other principles and reciprocity does not trump them. Obviously, justice and reciprocity have something to do with each other. Equally obviously, not every matter of justice is a matter of reciprocity.[15] Buchanan says, "[J]ustice as reciprocity can at best yield an account of what those who happen to be able to contribute in a given scheme of cooperation owe one another."[16] Evidently, Buchanan would not deny that there is something to the basic idea of returning value for

[14] Chapter 8 explains why I do not try to expand my promissory model of desert into a theory of punishment. I am unaware of any similarly weighty reason to reject reciprocity-based theories of punishment, but my arguments for reciprocity, so far as I can see, provide no basis for such a theory.

[15] Buchanan 1990, 244.

[16] Buchanan 1990, 238.

value. What is wrong is thinking our *only* obligation is to return value for value.

A similar theme is implicit in "Courtroom", the scene with which Part 3 began. In that scene, the judge is wrong to see reciprocity as even relevant, let alone as overriding other principles. Why? Because principles governing a court of justice are not principles of reciprocity. Reciprocity is king roughly within the context of the history of a personal relationship. In a courtroom, though, a judge's personal history should not be relevant. Likewise, when my department seeks to hire someone, I am obliged *not* to use my vote as a tool for returning favors. Some favors are not mine to give. Sometimes I have a duty to play a role substantially in service of purposes other than my own.

Monistic theories of justice are theories that try to reduce all of justice to a single principle. Pluralistic theories depict justice as consisting of several elements that cannot be reduced to one. A monistic theory of justice as reciprocity would have trouble explaining why reciprocity's domain is limited: Why, for example, it can be true that the judge in the "Courtroom" case really ought to return the favor he owes, yet also true that his courtroom is not the place. A monistic theory of justice as reciprocity would have trouble explaining obligations to those who did not help us – and therefore would have trouble explaining obligations to those who *cannot* help us. The problem, though, is the monism, not the reciprocity.

Buchanan is right to say our moral obligations go beyond our treatment of those who can relate to us on reciprocal terms. Whether torturing children is unjust does not turn at all (let alone only) on whether the children contributed (or ever will contribute) to cooperative ventures. However, Buchanan's conclusion should be not that reciprocity has no place in a theory of justice, but that reciprocity is not the only principle that has a place. Justice is not exhausted by principles of reciprocity, but reciprocity remains an essential thread in the fabric of a good community.

15

Varieties of Reciprocity

THESIS: As with desert, and justice more generally, reciprocity is a cluster concept.

RECIPROCITY'S CANONICAL FORM

Reciprocity in its canonical form applies to relations among autonomous adults, not to our dealings with those who are helpless. Crucially, if reciprocity pertains to the question of how to be a proper *recipient*, it will say nothing about what we owe to people who cannot do their share, because reciprocity concerns what we owe to people who have *already done* their share. With regard to people who did not (and perhaps cannot) do anything to put us in their debt, reciprocity does not tell us what to do. It does not tell us why. More generally, it says little about when or how or why to make the first move, because reciprocity primarily is about how to respond when someone else has made the first move.

We enter reciprocity's domain when we ask how to respond to people who have done us a favor. If that is not the question, then reciprocity is not the answer.

TRANSITIVE RECIPROCITY

Or at least, reciprocity in its canonical form is not the answer. There are variations on the theme, though, that speak to broader questions about moral obligation. Reciprocity in its canonical form obliges us to return favors specifically to our original benefactors. I refer to this form of reciprocity as *symmetrical* reciprocity.

Sometimes, the fitting response to a favor is not to return it but to "pass it on." When a teacher helps us, we are grateful. However, it is both odd and ordinary that we acknowledge debts to teachers mainly by passing on benefits to those whom we can help as our teachers helped us. I call this *transitive* reciprocity. Having received an unearned windfall, we are in debt. The moral scales are out of balance. The canonical way to restore a measure of balance is to return the favor to our benefactor, as per symmetrical reciprocity. However, the canonical way is not the only way. Another way is to pass the favor on, as per transitive reciprocity. Transitive reciprocity is less about *returning* a favor and more about honoring it – doing justice to it. Passing the favor on may not repay an original benefactor, but it can be a way of giving thanks. Professional athletes give money to college alma maters. While some seek literally to return a favor to the school, others seek to honor the school by working with it to benefit the next generation.[17] The intent is less to repay the favor than to pass it on.[18]

Both symmetrical and transitive reciprocity are cases of trying to do good in proportion to good received. Of course they are not the same, yet they are variations on this theme. Readers are welcome to reserve the sound "reciprocity" to refer to symmetrical reciprocity, using some other sound such as "schmeciprocity" to refer to what I call transitive reciprocity. It does not matter what sound we use, so long as we understand how they connect: Both are ways of trying to do good in proportion to good received.

In the symmetrical case, the favor is repaid. Sometimes, though, repaying is not possible. Other times, it would be inappropriate. If the original benefit is a gift, any attempt literally to repay runs a risk of dishonoring it. So, realizing this, we may instead pass it on, explicitly or implicitly in the original benefactor's name. In passing the favor on, we restore symmetry around ourselves, between what we have given and were given. If we cannot or should not try to restore symmetry around the original benefactor (between what the benefactor gave and was given), then transitive reciprocity may be the best we can do to balance the scale.

[17] Alumni who endow scholarship funds did not grow up with the younger generation of students who benefit from their philanthropy, but may feel kinship in virtue of imagining those students facing similar obstacles. I thank Walter Glannon for the thought.

[18] There also is such a thing as transitive *retribution* – taking revenge by punishing innocent people who share an offender's nationality or religion. Transitive retribution is as nightmarish as transitive reciprocity is inspiring.

I said duties of reciprocity inherently are vague. Intuitively, duties of transitive reciprocity possess all the earmarks of a Kantian imperfect duty. (That is, they tend to be indefinite regarding what one is to do, for whom one is to do it, and whether anyone has a right to enforce it.) However, although duties of transitive reciprocity are imperfect, there are limits to their indeterminacy. For example, if I feel indebted to my alma mater, giving a large gift to my alma mater's in-state rival may be a very good deed, but it is not a way of giving thanks to my alma mater.[19]

Some people feel indebted to society per se, and seek to give something back to society per se. We could call this a kind of symmetrical reciprocity, one naming society both as beneficiary and original benefactor. Sometimes, we would be unable to distinguish behaviorally between this and what I call transitive reciprocity. Bob and Hilda may each give blood to the Red Cross. Bob wants to repay the Red Cross (conceived as original benefactor) for saving his son. Hilda wants to pass on a life-saving favor to an anonymous beneficiary as a way of giving thanks for the anonymous donation that saved her daughter.[20] Hilda does not see herself as *repaying* the original benefactor (the anonymous donor) but does see herself as offering that donor a kind of salute. I do not suppose this is a major issue. Bob and Hilda themselves easily could miss the difference in their respective motives.

Still, when someone names a scholarship fund after a parent or former teacher, it is not a standard case of symmetrical reciprocity among trading partners. Neither is it simply a matter of joining in a collective effort to sustain a social structure (though it may be partly that). Nor is it a gesture of nebulous gratitude to society. It is a way of honoring a specific person.[21]

[19] Are acts of reciprocity successful only if they succeed in communicating to recipients (and/or the original benefactor) that the act is intended to acknowledge the receipt of a favor? I suppose not; I suppose we could take on a private mission to repay a debt to society, and could succeed in our own eyes without ever explaining our motivation to anyone. Still, I suppose we typically view our acts of reciprocity as successful only if they succeed in communicating that we are settling or at least acknowledging a debt.

[20] I present the example in terms of sons and daughters because if Bob and Hilda themselves had received transfusions, they would not be eligible to donate blood under current Red Cross rules.

[21] Becker (1986, 111) says that if the original gift "is aimed not at a specific person at all, but aimed instead at sustaining a social structure to provide such benefits to many people, we may conclude (in the absence of evidence to the contrary) that a return aimed at the same purpose will be fitting." Becker is right, and he is describing a subcategory of what I call transitive reciprocity. The larger category does not require that the original favor not be aimed at a specific person. When your teacher helped you years ago, transitive reciprocity can be called for even if the favor was aimed specifically at you. Even then, literally

Of course, we can and sometimes do feel a nebulous sense of gratitude, and it can help to motivate transitive reciprocity. I once was riding a bus in Vancouver, late at night. An Asian woman who did not speak English was trying to communicate to the driver that she was looking for a particular youth hostel but was lost. Having no luck with the driver, she was about to get off the bus on a deserted residential street in mid-winter at 10 PM with no idea what to do next.

I owed her nothing, or at least, nothing explicable in terms of symmetrical reciprocity. Yet, I saw the look on her face as she was about to get off, and it brought back painful memories of how it feels to be alone and lost in a place where I did not speak the language. So, I conveyed to her that I would get her where she wanted to go. We got off at a different stop, consulted a map, and walked to another stop where I put her on a different bus, making sure the driver would let her off at the hostel. The bus left with her bowing through the window and me feeling grateful. Grateful for what? Honestly, I am not sure. Grateful, I suppose, to be living in a mostly benevolent world, and grateful for the opportunity to thank the world for its benevolence by being a small part of what makes it benevolent.

RECIPROCITY AS A VALUE

Symmetrical reciprocity pertains to debts to our benefactors. Transitive reciprocity is a variation on this theme, pertaining to debts we acknowledge not by returning a favor but by passing it on. Both are ideas about indebtedness – about how reciprocity constrains us. What if we were to think of reciprocity as a value, and specifically as a value we should promote?

Suppose we see it as an open question whether our children will grow up as autonomous adults, willing and able to live as reciprocators. Suppose we think it would be *better* if they did – better for them, not only for their future neighbors and partners. In that case, we may consider ourselves obliged not to repay a debt but to promote a value. Specifically, we promote a capacity to repay debts, and we nurture the kind of character that takes joy in putting that capacity to use.[22]

returning the favor can be infeasible or inappropriate. Thus, even then, sometimes the most fitting response to a favor is to pass it on.

[22] We value reciprocation per se, but what we *promote* is the willingness and ability to reciprocate, because at some point free will has to take over. I thank Michael Smith for this point.

When we see reciprocity that way, we see ourselves as owing the next generation something other than simple reciprocation, something having nothing to do with what children have done for us and everything to do with who they can become with our help.[23] As children grow up, they acquire a duty to live as responsible adults, which includes (for one thing) reciprocating. In virtue of that pending obligation, we try to put children in a position honorably to meet that obligation.

Buchanan's main charge against justice as reciprocity is that anyone who thinks membership in a moral community is restricted to those who can reciprocate is taking reciprocity too seriously. Buchanan has a point. We may add a complementary point: What Buchanan calls justice as reciprocity also fails to take reciprocity seriously *enough*. We are born unable to reciprocate. Thus, since reciprocity is at the core of a viable community, a moral community will not take the capacity to reciprocate for granted. It will *work* to enable people to live as reciprocators, while moral citizens accept some individual responsibility for fostering their community as an enabler of reciprocity.

We normally envision reciprocity entering moral deliberation as a constraint – as a duty to return favors. Reciprocity as a value, though, functions differently in moral deliberation. Reciprocity as a value is a goal, not a constraint.

ALIENATION, RECIPROCITY, AND THE IDEAL OF EQUAL PARTNERSHIP

Becker says that if act A is not perceived as a *good*, as a *return*, and perceived as such in time to *affect future interaction*, then act A will have no point as an act of reciprocity.[24] This is overstated, but Becker does have a point. Becker's point is that without these features, acts of reciprocity cannot have reciprocity's canonical instrumental value, which is to help us to build ongoing mutually advantageous relationships. It is an overstatement, though, to suggest that if reciprocity lacks *this* point, it has *no* point. Reciprocity has purposes other than the canonical one. Reciprocating need not always be done with a view to facilitating future interaction. We might return a favor for the simple reason that we owe a favor. Or,

[23] With children, we act on behalf of what they *can* become. With adult beneficiaries (when the point is to promote reciprocity as a value) we intuitively set a higher standard; that is, we condition duties on what a person *will* become, or realistically *intends* to become.

[24] Becker 1986, 107.

we might return a favor as a way of acknowledging and celebrating our common humanity.

The passages in Marx I find most illuminating are his remarks on alienation.[25] I count myself lucky to live in the kind of society of which Marx is a foremost critic, yet I still find his remarks on alienation to be accurate. Liberal society does alienate some people: Their work, their selves, their possessions, their species, and even nature itself confront some people as alien forces. Liberal society may be less alienating than critics say, and palpably less alienating than realistic alternatives, in part because it lets people associate on their own terms. It lets people create "thicker" communities. But this does not mean it is not alienating.

Perhaps alienation is a permanent part of the human condition. Perhaps there is no cure, or at least, no large-scale, permanent cure. Nevertheless, we can solve the problem on a personal scale. On a personal scale, failing to reciprocate is among the most alienating things we can do. (Reflect for a moment on the fact that failing to *accept* favors can likewise be alienating.) When we fail to respond, we cut ourselves off not only from mundane benefits of mutual support but also from relations that make us feel visible and valuable. My dentist once did a bit of work for me and, for reasons I do not fully understand, declined to charge me for the service. It would have been wrong to respond by sending my dentist a check, but also wrong if I had not sent a thank you card or otherwise expressed my gratitude. The issue is not simply prudence. Rather, it concerns what people like my dentist and me need to do to sustain a vivid picture of ourselves as rightly esteemed agents in a world of rightly esteemed agents.

At best, squandering an opportunity to reciprocate squanders a chance for mutual affirmation – to affirm that our partner was right to see us as worthy of her trust. Reciprocating shows our partners that we value them as ends in themselves. More implicitly, but still obviously, honoring those who treat us as ends shows that we value *ourselves* as ends.

Accordingly, I do not believe that reciprocity and gratitude are called for only in response to people who are going beyond the call of duty. Because reciprocity and gratitude are forms of mutual affirmation, it makes perfect sense to feel grateful to people simply for doing their duty. If I notice when motorists do the right thing in heavy traffic, and give a wave of appreciation, I make the road a safer, more courteous place.

[25] For example, see Easton and Guddat's collection of essays by the young Marx (1967, 287–301).

If I thank cashiers for giving me good service – not beyond the call of duty but good enough – I make that store a better place to work and shop.

Normal competence is an achievement, not an effortless default. Rising to the call of duty is an achievement; it is not a mistake to give people credit for doing so. Even if the ordinary, expected response to ordinary, expected behavior is to show no appreciation, there is no mistake in exceeding the norm. Reciprocators do not avoid giving credit; they seek not to pay as little as possible for maximally valuable favors but rather to acknowledge gladly the roles they and their partners play in cooperative ventures.[26]

Therein lies an element of equal respect and equal treatment that is part of the essence of being a reciprocator. Moreover, unequal though people may be along a given dimension, they can devise terms of interaction that emphasize dimensions along which they have the most to offer each other. As reciprocators, people *craft* dimensions along which they can relate as equals. Such crafting is not a panacea for society's ills, but reciprocators can and do address their personal alienation problem, one relationship at a time.

Reciprocity is antithetical to the "atomism" that liberalism's critics inaccurately say is characteristic of liberal society.[27] Reciprocators know that a transaction that goes well is a mutual affirmation, and therefore that participating in such transactions is self-affirming. Alienation is fundamentally a personal problem. The antidote is active affirmation of our common humanity – the kind of affirmation we practice when we practice reciprocity. Transitive reciprocity, in particular, is a way of rejoicing in our common humanity. None of this is mere "second best."

Two caveats. First, as reciprocity reinforces solidarity within a group, it may exacerbate between-group alienation, even while some individuals ameliorate the between-group problem by building personal relations with members of other groups. It is agents, not classes, who reciprocate, so reciprocity breaks down class barriers on a person by person basis, not in some more general way. Second, market transactions obviously need

[26] I do not mean here to deny Gauthier's insight that being a "constrained maximizer" can be a way of securing maximum personal gain. Under certain circumstances, operating under constraints is a way of being the sort of player that people seek out as a trading partner, whereas "straightforward" maximizing is a way of being the sort of player whom others shun.

[27] For a seminal discussion of the problem of atomism, see Taylor 1995, 139ff.

not live up to the ideal of mutual affirmation. Customers and cashiers who do not look each other in the eye are not helping each other to feel less alienated. However, a customer who makes a point of noticing when shopkeepers are good at what they do does indeed make the world a less alienated place. Money is seldom the only thing shopkeepers want from their customers.

16

Debts to Society and Double Counting

THESIS: Persons are not in debt simply in virtue of being persons in a society. If they are in debt, it will be in virtue of their unique individual histories.

THE DOUBLE COUNTING PROBLEM

A principle of reciprocity sometimes is invoked to ground political obligation (especially an obligation to pay taxes). In the hands of communitarians and nationalists, such arguments begin with the fact that, as Lawrence Becker says, "No one is self-made. Whatever good there is in our lives is, in part, a product of the acts of others."[28] What is Becker getting at? We might think Becker is about to conclude that while we owe society for what it did for us, we deserve no credit for what we did in return. Not so. In fact, Becker has this to say, in the form of a parable:

> I know a man who thinks he lives in debt. . . .
>
> He acts as though he were never, ever, fully entitled to anything. As though good will, and good motives, and conscientiousness on the part of others were never his by right or reason, but always something to admire.
>
> He is grateful to his parents, even though he cares for them in their old age. He is grateful to his employer, and loyal, even though his work is barely noticed. And he is grateful to his country too (the strength of those emotions is embarrassing), even though he's suffered for it.
>
> He is aware of what he does for others. He just thinks that everything he is and has is somehow owed to them. Without them, he would never have been anything. Without them, he would collapse like a house of sticks.

[28] Becker 1980a, 9. See also Becker 1980b, 414.

> It is, he thinks, a debt that cannot be repaid. He is not, he thinks, an atom – not a solitary, lonely, self-sufficient provider of his own good life. . . .
>
> He is a fool.[29]

Becker cautions us not to assume he stands exactly where his narrator does in such parables (which are scattered throughout his book). So Becker may not be so sure the man in the story is a fool. Becker may agree there is nothing foolish about feeling grateful "all the way down." What is foolish is failing to see that the debt is limited and not beyond repayment. It may be foolish to see ourselves in "atomistic" terms, but it is neither "atomistic" nor foolish to think it is possible to have done our share. Note: Feeling we have done our share is not the same as feeling ungrateful. We can have reason to feel *grateful*, and to want to do more, even when we have no reason to feel literally *indebted*.

Needless to say, in addition to benefits we receive from specific trades, we benefit from living in a productive, prosperous society. Is this an extra favor? If so, what would repay it? Have we already repaid by participating in the same trades through which our partners receive (at the same time as they, like you, help to create) those very same background benefits? Might we still have some amorphous duty to repay society even after every member of society has (just like us) paid in full for services rendered?

My point is not that answers to these questions are obvious. Instead, I am saying that however we answer these questions, we need to be careful, in our search for foundations for obligations to society, to resist the temptation to double count. It is not as if a worker who shows up for work every day and does the best she has so far only been *taking*. We need to keep in mind that, just as the trade of millions of people adds value to our lives, so too does our trade add value to millions of lives in return. It would be a mistake to discount either side of this equation. Jane's discounting what others contribute to her life is bad, but no worse than our discounting what she contributes to the lives of others in return; both are instances of the same mistake.

We sometimes speak as if the only way to "give back" to society is by paying taxes, but any decent mechanic does more for society by fixing cars than by paying taxes. Or if that is not obvious, then consider Thomas Edison. There is no amount of money that Edison could have paid in taxes that would have begun to compare to what Edison contributed to

[29] Becker 1986, 6.

society when he gave us the light bulb. We gave Edison a fortune, but what we gave Edison was nothing compared to what Edison gave us.

IF IT'S YOUR TALENT, SHOULD YOU GET ALL THE BENEFITS?

I sometimes am asked why I think talented people ought to get *all* the benefits that accrue from exercising their talents, but of course, I don't think that. What I think is that when talented people bring their talents to market, far-reaching ripples of mutual benefit are set in motion. Life expectancy nearly *doubled* in the course of a single, tragedy-filled, yet technologically (and in some ways culturally) progressive century, and for all we know may continue to rise in the century to come. Free societies make progress. They do not make progress without people such as Thomas Edison (whose light bulbs surely contributed to the increase in life expectancy) but on the other hand, free societies tend to produce people like Edison.

Jane normally has to bring something to the market before she can take anything away. Not everyone likes this fact about markets, but it would be inconsistent to note this fact and then go on to say Jane ought to give something back, as if she had not already given. This is not to say Jane would be mistaken to think about her legacy, and thus about how she might give back even more. The mistake would lie in thinking that if Jane is better off, she must not yet have done her share.[30]

If Jane participates in networks of mutual benefit, then by that very fact, she is more or less doing her share to constitute and sustain those networks. Admittedly, if Jane receives an average reward in a society like ours, she receives a package of staggering value (more than even Edison could have imagined as recently as a century ago). The fact that everyone's doing a little results in huge gains for nearly all makes it right for Jane to feel grateful to be a part of the enterprise. Still, if everyone is doing a little, then doing a little is Jane's share. Obviously, an average person *gets* a lot, but this does not determine what counts as an average share of the *doing*.[31]

[30] Actually, Jane can enjoy some goods – public goods – without bringing anything to market, but these too are cases where if we get specific about what to count as Jane's share, Jane may have already done it. If Jane's neighbors put up Christmas decorations, Jane is enriched. But if Jane already put up comparable decorations, then she already did her share.

[31] Jane may owe more than that, but her extra debts evidently would not be matters of reciprocity. Or, perhaps there is a difference between first- and third-person perspectives

CONCLUSION

This conclusion is controversial, by no means widely accepted, so I will make the point again, in different words. If society would be better off without Jane, then we have some reason to say Jane has an unpaid debt. Jane has not been "carrying her weight." But if Jane already has contributed enough to make society better off by virtue of Jane being part of it, then there is no basis for saying Jane has an unpaid debt. That *Jane* is that much better off for participating in society is not what determines whether she is in debt. Whether she is in debt depends on whether *society* is that much better off in turn.[32]

DISCUSSION

We are better off with society than without it. We also are better off with the sun than without it. So what? Is the sun doing us a favor? Is society?

on what Jane owes. That is, even if every favor ever done for Jane were repaid (not necessarily by Jane), it can still make sense for Jane to feel so grateful that she chooses to react as if she were in a state of literal indebtedness. Such an attitude is common among philanthropists, and I see no reason to question it.

[32] I left open whether collective entities can give and receive favors. Some collective entities – corporations, for example – are agents, or like agents, in crucial ways. See Rovane 1998. Societies, though, are barely like entities at all, let alone like intentional entities. Maybe a society is enough like an entity that it can be done a favor, but not enough like an agent to do a favor.

17

The Limits of Reciprocity

THESIS: No single principle is more than an element of justice. Principles of reciprocity, though, are at the core of a just society.

RECIPROCITY AND BASIC STRUCTURE

Allen Buchanan says, "To the extent that justice as reciprocity is conceptually barred from even considering the justice or injustice of the choice of the fundamental framework for cooperation and hence the choice of criteria for membership in the class of contributors, it is a superficial view. Indeed, justice as reciprocity is incapable of even recognizing what may be the greatest injustice a person can suffer: morally arbitrary exclusion from the class of subjects of justice."[33]

First, a preliminary: What Buchanan calls superficial is not such a bad thing. A basic structure is not the only entity that has a job to do. It would be foolish to think our society is just only if its basic structure all by itself guarantees that we all get whatever we are due. Much of what we are due comes to us via each other, not via basic structure. Inevitably, citizens handle (and must be trusted to handle) much of the responsibility for making sure they and their neighbors are, within reason, treated fairly. Not every matter of justice is a matter of basic structure.[34]

Preliminaries aside, though, I agree with Buchanan that it would be odd for a theory to have nothing to say about basic structures that form the background against which everyday reciprocal exchange takes place.

[33] Buchanan 1990, 239.
[34] On this topic, see especially Tomasi 2001, chap. 6.

On one hand, the people who carry on from where they are, in win-win fashion, make society a cooperative venture for mutual advantage. Yet, it would be crazy to interpret this fact as implying that whether it is right to abolish slavery depends (case by case?) on whether abolition is a win-win move for slaves and slaveowners alike.

Of course, no one says that. In fact, justice as reciprocity need not be so cartoonish a notion. First, reciprocity, even symmetrical reciprocity, implies no such thing. Symmetrical reciprocity concerns what people owe to their benefactors. It is silent on questions of basic structure, correctly recognizing that reciprocity (especially reciprocity as a constraint) is but one subdomain of the larger realm of justice. Symmetrical reciprocity does not pretend to explain what is wrong with slavery.

Second, reciprocity as a value *is* relevant to evaluating basic structure. It can answer questions that symmetrical reciprocity cannot, perhaps even including questions about what is wrong with slavery. Using a conception of reciprocity as a value, we can assess basic structure by asking whether it fosters norms of reciprocity; whether it puts people in a position to deal with each other as autonomous adults, in mutually beneficial ways; whether it puts people in a position to deal with each other in ways that make them glad to have each other as neighbors.[35]

As Buchanan stresses, the ability to contribute is in part socially determined. We are not self-made. I agree. To be sure, modern technology enables those who lack brute strength (women in particular) to contribute vastly more to the economy, and on a more equal – a more reciprocal – footing than would have been imaginable a few generations ago. So what? Obviously, it does not follow that we should discount contributions of nonself-made people. On the contrary, if the idea that we are not self-made has any practical point, the point is not that we should give women (or men) less credit for what they contribute, but that we should applaud our society for putting women (and men) in a position to contribute as much as they do.

Thus, if Buchanan is right that social structures play a role in enabling people to contribute, this *establishes* rather than precludes the possibility of our assessing and sometimes condemning social structures

[35] I do not mean to say reciprocity as a value can ground a complete or foundational account of what is wrong with slavery, only that it is on topic in a way that symmetrical reciprocity is not. The ideas that people deserve a chance (Chapter 8), that people command equal treatment (Chapter 19), and that we are separate persons with separate entitlements (Part 6) would be among the ideas that would figure in a more complete account.

(including slavery) in terms of whether they play this role well. A theory incorporating reciprocity as a *value*, then, has plenty to say about social structure.

Here is a crucial contrast between reciprocity qua constraint and qua value: If we apply reciprocity qua constraint to social structure, we may conclude that a social structure should distribute benefits of cooperation in proportion to what people contribute. By contrast, the pattern prescribed by reciprocity qua *value* is not distribution in proportion to contribution, but distribution to produce citizens willing and able to reciprocate – to participate in the patterns of mutual recognition.

THE DISABLED

Some people return favors more easily than others, so we may wonder whether those who find it more difficult are in a different position regarding duties to return favors. And surely they sometimes are, although it may be hard to specify the exact nature of the difference. (Imagine a normally ambulatory person holding open a door for a person in a wheelchair, and the person in the wheelchair wanting to reciprocate by holding open the next door. The person who *owes* the favor may believe the extra difficulty makes no difference at all, whereas the original benefactor may be mortified at the thought of the beneficiary going to all that trouble, and may wish the beneficiary would simply say thanks and leave it at that.) In any case, the topic of what we owe each other is not exhausted by discussion of reciprocity, and this fact is especially relevant when we discuss obligations to the disabled.

Some of the skepticism about reciprocity stems from worries that a society grounded in an ethos of reciprocity would be prejudicial to the disabled; that is, as a criterion for moral standing, it would exclude the disabled. Quite the contrary: In practice an ethos of reciprocity tends to *help* disabled people to participate in their communities. This is so for two reasons.

First, many disabilities are only contingently incapacitating. When norms of reciprocity function as engines of social and technological progress, they enable people to live as autonomous reciprocators whose disabilities otherwise would have been incapacitating. For example, I am terribly nearsighted. However, because I can buy glasses at any shopping mall, no one even thinks to classify me as disabled. (As I write, 'disabled' is the currently label, but it is misleading, referring as it does to problems that are only contingently disabling.) In short, when all goes well, reciprocity empirically tends to foster conditions in which people like me do not need carrying.

Second, an ethos of reciprocity tends to foster conditions in which, when our disabled neighbors are truly in need, the rest of us can afford to carry them. For these two reasons, disabled people have a practical stake in fostering society as a reciprocal venture, *even if* the concept of reciprocity is irrelevant to the theoretical task of grounding special obligations to them.

Reciprocity does not ask us to carry our neighbors, but that is not reciprocity's job. If carrying those with disabilities is a matter of justice, it is a different element of a pluralistic theory that yields such duties. But this observation takes nothing away from the principle of reciprocity within its own domain, even from the perspective of the disabled.[36] To summarize, if reciprocity is not the element of justice that yields special obligations to the disabled, it may yet sustain an economy and a culture in which mildly to moderately disabled people can live more or less normal lives, and moderately to severely disabled people can get special care when they need it.

FAVORS ACCEPTED VERSUS FAVORS MERELY RECEIVED

When do I acquire a duty to return a favor? When I *accept* the favor? Or do I acquire the duty simply in virtue of *receiving* the favor, whether or not I willingly accept it as such? Robert Nozick imagines a case where your neighbors institute a public address system. "They post a list of names, one for each day, yours among them. On his assigned day (one can easily switch days) a person is to run the public address system, play records over it, give news bulletins, tell amusing stories he has heard, and so on. After 138 days on which each person has done his part, your day arrives. Are you obligated to take your turn?"[37]

Surely not, Nozick says. The idea of neighbors unilaterally deciding to do something for you, then unilaterally deciding what you owe in return, is outrageous. On the other hand, although we may agree that

[36] Becker, significantly disabled himself, reflects on such issues in (1998).

[37] Nozick 1974, 93. In a discussion of Nozick's story, Simmons (1979, chap. 5) notes that it matters whether the venture is worth its cost to a prospective contributor. More oddly, if a listener does benefit, then, intuitively, it somehow matters whether the listener lives in the neighborhood or merely works there during the day.

our neighbors have no right unilaterally to decide what we owe them, that implies at most that we bear some responsibility for deciding how to respond. It does *not* imply that the correct response is for us to decide we owe nothing. (What if the good at stake is not a public address system but something more uncontestably good and important, such as a flash flood lookout? If the good were compellingly important, would you then be obligated to take your turn, even if you did not consent to do your part in providing it?)[38]

We receive unasked-for favors all the time. A proper moral agent does not enjoy them with no thought of reciprocation. Proper neighbors, in turn, acknowledge that, if the question truly concerns reciprocity (as opposed to, say, the adequacy of voluntary public goods provision), then recipients must decide how to respond. Philosophers are left with a puzzle. To acquire a duty to reciprocate, must you accept a favor as such, or does merely receiving it suffice? Intuitively, merely receiving is not always sufficient, but full-blown acceptance is not always necessary.[39]

Speculating in reciprocity involves doing favors without asking – not giving recipients a chance to decline – in order to obligate them.[40] If our neighbors speculate, they bear the risk of being disappointed with what we deem a fitting response. If they are not speculating but simply want to benefit us, then so long as we do in fact benefit, they will not be

38 Becker's claim that reciprocity is a virtue that generates nonenforceable duties is analogous to Wellman's (1999) claim that gratitude is a virtue but not a duty. See also McConnell 1993. Simmons (1979, chap. 7) distinguishes two obligations of reciprocation: fair play and gratitude. The distinction arguably would matter if we were trying to ground specifically political obligation, although Simmons himself concludes that neither kind of grounding can work.

39 Between passively receiving and willingly accepting is a middle ground. What if you *would have* accepted F had it been offered? A doctor performs emergency surgery on a comatose accident victim. The patient regains consciousness, and the doctor says, "I had to assume you would've consented if I had been able to ask." If the doctor is correct, it seems the patient owes the doctor a favor. Instead, suppose the patient had wanted to die and would not have consented. However, that patient, previously suicidal, is now thrilled to be alive, and thus has received a true favor to which she would not have consented. If, in that case, the patient owes the doctor a favor, then hypothetical consent is not always necessary. It is widely acknowledged that hypothetical consent is not enough, either, at least when trying to ground enforceable political obligations. Thus, this middle ground proposal does not solve our puzzle, but I thank Steve Biggs for the suggestion. I also thank Nick Sturgeon for (in conversation) responding to an article of mine on hypothetical consent (Schmidtz 1990a) by correctly suggesting the emergency surgery case as a paradigm case in which hypothetical consent would be meaningful.

40 The idea of "speculating" comes from Becker 1980b, 419.

disappointed. Or, if they are traders, not speculators, they will negotiate beforehand, if they can, to avoid misunderstanding about what would be a fitting response.

Becker says we acquire no debt to speculators because speculators do not benefit us.[41] This may be a case of getting the right conclusion for the wrong reason. Another reason why we owe no favor to speculators, even when they do benefit us, is that the thought counts.[42] And the thought behind speculation is not benign. A speculator's intent is not to do us a favor but to invest in our propensity to feel obligated. Becker and Nozick could agree that if the thought is not benign, we acquire a debt only if we willingly accept the favor. Speculative favors, then, arguably are a category of favors for which involuntarily receiving the favor is not enough.

REAL FAVORS, AND HOW WE RETURN THEM

Speculative favors are favors done for the purpose of obligating beneficiaries. Sometimes, theorists seeking to ground political obligation will, in effect, look for something they can *call* a favor, and appeal to that as a basis for grounding obligations to pay taxes, serve in the military, and so on. Needless to say, we have an obligation to return a favor only if a favor actually was done. If some identifiable party did us a favor, then that party is the party to whom we will owe a favor. Arguably, we all owe something to Thomas Edison for inventing the light bulb. However, if Thomas Edison already has been paid in full for his invention, it would seem we are too late – the favor has already been returned, more or less.

Perhaps we owe something to people who paid Thomas Edison, thereby discharging what otherwise might have been our debt to Edison? If we say that, we leave the situation where we owe a favor to some identifiable party like Edison. The obligation morphs into a nebulous debt to society. We owe a favor to everyone who ever bought a light bulb, or worked for a company that produced light bulbs, or helped to bring light bulbs to market and sell them to customers like us.

There is a grain of truth to this idea. What then do we owe such people? Presumably, what we owe them is to *return the favor*. We ought, like them,

[41] Ibid.

[42] There is a saying "It's the thought that counts." I see no reason to believe *only* the thought counts, but the thought does count. We want to be acknowledged.

to pay for light bulbs and whatever else we consume, and ought to earn our income by working for firms that produce or market light bulbs or otherwise make people better off. When we do that, we respond in kind to what they did for us. The favor they did was to earn a living, and we are earning a living in return.[43]

WHEN ARE DUTIES OF RECIPROCITY ENFORCEABLE?

Nozick identified, and Becker gave a name to, the problem of speculative favors. Becker may also have a partial solution: If our neighbors are speculating, our merely receiving their favors does not oblige us to reciprocate. That leaves open questions about what we owe to neighbors who are not speculating but who instead are sincerely trying collectively to provide a public good.

We could leave such questions open, or classify them as matters of personal conscience. Unless we seek to translate duties of reciprocity into a justification of coercion, we can accept the plain fact that duties of reciprocity are imperfect. We receive benefits from diffuse and dispersed sources. Intuitively, we should feel grateful, and should be disposed to be part of the network of benevolence that makes people in general better off. Something goes wrong if we do not gladly join that network. Something equally goes wrong if other people force us to participate against our will. Pulled as we are by conflicting intuitions, we do not jail people for failing to return favors. At most, we stop trusting the person, withdraw from the relationship, or perhaps gossip. We do not turn the other cheek, but neither do we resort to official channels in search of retribution. In Adam Smith's words, "To oblige [a person] by force to perform what in gratitude he ought to perform, and what every impartial spectator would approve of him performing, would, if possible, be still more improper than his neglecting to perform it."[44] The concept of reciprocity is thus a decidedly awkward model for any sort of enforceable political obligation.

Therein lies part of the point of contract law. Reciprocity in its canonical form is not enforceable. Part of the point of contract law is to convert

[43] Becker (2003) recently quipped, "living in a political system we did not choose and cannot leave, a system which showers us with things like Terminator movies, does not by itself create obligations of reciprocity in us."

[44] Smith 1982, 79.

the obligation to a form that is enforceable. Parties who enter into a contract are licensing their partners to rely on them to do what they have consented to do, as part of a reciprocally beneficial transaction. They also are licensing their partners to seek compensation if they should fail to perform as promised. It ought to go without saying, though, that if there is no actual contract, then there is no actual licensing, either. Hypothetical consent models of political obligation in this way fail to take contracts seriously, thus failing to take consent seriously, thus failing rather profoundly to take reciprocity seriously, too.

My hypothesis is that if we reject Nozick's conclusion that he does not owe his neighbors a day at the microphone, it is not because we care more than Nozick does about reciprocity. The real reason is, we worry more than Nozick does about public goods. Our greatest worry, on this view, is not to get the right *input* from each person so much as to get the right *output* from the collective effort as a whole. Many would say some public goods (for example, military service) are too crucial for us to trust in voluntary participation. We simply must have a guarantee.

Thus, talk of reciprocity distracts us from social and political concerns that give us real reasons to consider employing coercion. If we have any good reason to coerce people, it will not be because we cannot stand the thought of someone somewhere failing to return a favor. A good reason will be something else: For example, perhaps we cannot stand the risk of inadequate public goods provision (and we at least hope that if government taxes us, the risk will be lower).

SUMMARY

Buchanan worries that reciprocity as a principle of justice justifies too little. I replied that it is not reciprocity's job, within a pluralistic theory, to justify everything worth justifying. Moreover, reciprocity has dimensions beyond those considered by Buchanan, so reciprocity can justify more than Buchanan suggests. If it cannot justify everything, that is not a problem. I agree with Buchanan that what he calls justice as reciprocity is inadequate as a theory of justice. However, I argued, what he calls justice as reciprocity also is inadequate as a theory of reciprocity. A better theory of reciprocity would play a prominent role in a plausible theory of justice.

1. How should autonomous equals operating at arm's length treat each other? How should partners in intimate relationships treat

each other? How should fully capable adults treat people – for example, elderly parents – who made their contribution before becoming permanently disabled? Part of the answer to all three questions is that people should return favors, as per *symmetrical* reciprocity.

2. What if we want to acknowledge debts, but literally repaying would be inappropriate or infeasible? Part of the answer: There are times when we should pass it on, as per *transitive* reciprocity.
3. How should fully capable adults treat people who cannot yet contribute, especially children? Part of the answer is, fully capable adults should embrace reciprocity as a *value*, acting on behalf of the autonomous but mutually supportive reciprocators that their beneficiaries may yet become.
4. How should fully capable adults treat the profoundly disabled – people who have never been and never will be even minimally capable adults, people who never have and never will do a meaningful favor for anyone? These questions are not answered by appeal to reciprocity. If we must answer such questions with a principle of justice, it will have to be some other part of a larger theory. Yet, although reciprocity does not answer such questions, it can affect how often such questions arise. In a just society, various forces work over time to reduce the extent to which disabilities (such as my poor vision) are contingently incapacitating, thereby reducing the number of people who fall outside the scope of reciprocity. Reciprocal trade as an engine of prosperity and progress is one such force. Reciprocal trade also helps to make people prosperous enough to have meaningful conversations about how much more justice requires them to do for their profoundly disabled neighbors.

There is a place in society for basic structures that work to reward people in proportion to what they have contributed to society. There also is a place, of equal or greater importance, for structures that work to put people in a position where they are willing and able to live as reciprocators. I conclude there is a prominent place within a theory of justice for traditional commonsense principles of reciprocity.

No single principle of justice, including justice as reciprocity, is more than an element of justice. Still, reciprocity is at the core of a just society, and needs to have a corresponding place at the core of our theories.[45] Relations of reciprocity are ultimate exemplars of the ideal of society as a cooperative venture for mutual advantage.

[45] Becker concludes that we should consider rejecting any "effort to construct elaborate conceptions of justice that fail to develop equally elaborate conceptions of reciprocity. A case in point, as I have mentioned, is Rawls's theory of justice, which over the years came to rely more and more on references to reciprocity but which, as far as I can tell, never seriously tackled the problem of getting a good general conception of it" (2003, 12).

PART 4

EQUAL RESPECT AND EQUAL SHARES

18

Equality

"Hey honey, aren't you enjoying Billy's birthday party?"
"Daddy, how come Billy got a bike and I didn't?"
"Oh, Cindy. Let me give you a hug. It's Billy's eighth birthday, honey. You'll get a bike too, just like his, on your eighth birthday. I promise. But you're only six years old. You have to wait a bit."
Cindy pushed away. "Daddy, you're supposed to treat us the same. If you give Billy a bike, you give me a bike. And if he gets his now, I get mine now."[1]

We are all equal, sort of. We are not equal in terms of our physical or mental capacities. Morally speaking, we are not all equally good. Evidently, if we are equal, it is not by virtue of our actual characteristics but despite them.

In fact, what we mean by equality is best seen as political rather than metaphysical (or even moral) in nature. We do not expect people to be the same, but we see differences as having no bearing on how people ought to be treated as citizens. Or differences, when they do matter, will not matter in the sense of being a basis for hierarchical class structure. People once saw society as consisting of separate classes – commoners and people of noble birth – but that belief belongs to another age. As a society, we made moral progress. Such progress consists in part of progress toward political and cultural equality.

At a certain stage in his argument, Rawls treats as "morally arbitrary" whatever makes people unequal, and indeed, whatever makes people nonidentical. Fair bargainers are supposed to ignore their different

[1] I thank Gerry Mackie for the story.

temperaments, talents and, most strikingly, that some people have *done* more than others.[2] So, it is no surprise that Rawls's conclusion has an egalitarian flavor. What *is* surprising is that the conclusion is not a strict form of egalitarianism. Having designed an original position that sets aside *everything* that makes people unequal, the door remains wide open for unequal shares. This may prove to be Rawls's lasting contribution.

If we understand Rawls in this way, Part 4's aim is complementary to Rawls's. Where Rawls argues that even egalitarians should embrace some forms of inequality (those that benefit the less advantaged), I argue that even nonegalitarians should embrace some forms of equality. Even for beings as unequal as humans are, there remains room within a pluralistic theory of justice for egalitarian elements.

Chapter 19 argues that there is a (limited) place for distribution according to a principle of "equal shares" even in otherwise nonegalitarian theories of justice. Nevertheless, "equal shares" is only one way of expressing egalitarian concern. Chapters 20 and 21 connect egalitarianism to humanitarianism and meritocracy, respectively. Chapter 22 explores the tricky empirical question of whether we are moving toward or away from a regime of equal opportunity. Chapter 23 considers and rejects the claim that there is a utilitarian argument, grounded in the concept of diminishing marginal utility, for a regime of equal shares. Chapter 24 reflects on why rules of first possession limit attempts to distribute according to principles (not only egalitarian principles) of justice.

[2] Rawls 1971, 72.

19

Does Equal Treatment Imply Equal Shares?

THESIS: There is a deep connection between equal treatment and justice, but not between equal treatment and equal shares.

ON BEHALF OF EQUAL SHARES

Bruce Ackerman's essay, "On Getting What We Don't Deserve," is a dialogue that beautifully captures the essence of egalitarian concern about differences in wealth and income. Ackerman imagines you and he are in a garden.[3] You see two apples on a tree and swallow them in one gulp while an amazed Ackerman looks on. Ackerman then asks you, as one human being to another: Shouldn't I have gotten one of those apples?

Should he? Why? Why only one? What grounds our admittedly compelling intuition that Ackerman should have gotten one – exactly one – of those apples? Notably, Ackerman denies that his claim is based on need, signaling that his concern is not humanitarian. Instead, Ackerman's point is that one apple would have been an equal share. To Ackerman, the rule of equal shares is a moral default. Morally, distribution by equal shares is what we automatically go to if we cannot justify anything else. In Ackerman's garden, at least, to say Ackerman does not presumptively command an equal share is to say he does not command respect.

Is Ackerman right? Looking at the question dispassionately, there are several things to say on behalf of "equal shares" *even if* we reject Ackerman's presumption in favor of it. In Ackerman's garden, equal

[3] Ackerman 1983, 60–70. For an egalitarian critique of Ackerman's "thought experiment" method, see Galbraith 2000: 387–404.

shares requires no further debate about who gets the bigger share. No one envies anyone else's share. When we arrive all at once, equal shares is a cooperative, mutually advantageous, mutually respectful departure from the status quo (in which none of us yet has a share of the good to be distributed). Finally, if we each want both apples, "splitting the difference" is easy, and often it is a pleasant way of solving our distributional problem. In the process, we not only solve the problem but offer each other a kind of salute. In Ackerman's garden, equal shares is an obvious way to get on with our lives with no hard feelings at worst, and at best with a sense of having been honored by honorable people.

These ideas may not be equality's foundation, but they are among equality's virtues. Crucially, even nonegalitarians can appreciate that they are virtues. Thus, even critics of egalitarianism can agree that there is a place in a just society for dividing some goods into equal shares. In particular, in "manna from heaven" cases, when we arrive at the bargaining table at the same time, aiming to divide goods to which no one has made a prior claim, we have a situation where equal shares is, from any perspective, a way of achieving a just distribution. It may not be the only way. (For example, we could flesh out the thought experiment so as to make bargainers' unequal needs more salient than their equality as citizens.) But it is one way.

DIMENSIONS OF EQUALITY

Amartya Sen says we are all egalitarians in a way, since "every normative theory or social arrangement that has at all stood the test of time seems to demand equality of *something*."[4] By the same token every theory demands inequality too, including egalitarian theories. An egalitarian is someone who embraces one kind of unequal treatment as the price of securing equality of (what he or she considers) a more important kind. (Recall Chapter 3's discussion of how tax cuts disproportionately benefit those who were, after all, paying the taxes.)

Suppose an employer routinely expects more work, or better work, from Jane than from Joe but sees no reason to pay them differently. The problem is not wage differentials (by hypothesis their wages are the same) so much as a lack of proportion between contribution and compensation. The lack of proportion is a kind of unequal treatment. And unequal treatment, and the lack of respect it signals, is what people resent.

[4] Sen 1992, 12.

Children often are jealous when comparing their shares to those of siblings: more precisely, when comparing shares doled out by their parents. Why? Because getting a lesser share from their parents signals that they are held in lower esteem. They are not so upset about getting less than their rich neighbor, because so long as no one is *deliberately assigning* lesser shares, no one is sending a signal of lesser esteem. Here, too, the problematic departure is from equal treatment rather than from equal shares.

Notice: As children grow up, we expect them to resent siblings less rather than to resent neighbors more. Resenting siblings less is a sign of maturity; resenting neighbors more is not.

NONSIMULTANEOUS ARRIVAL

Unequal treatment presupposes treatment. Unequal shares does not. When Ackerman is being *treated* unequally, there is someone whom Ackerman can ask to justify treating him unequally. Moreover, in Ackerman's garden, your grabbing both apples arguably is a token of unequal treatment.

What if Ackerman arrives years later, long after you have turned those two apples into a thriving orchard? Do you owe Ackerman anything? If so, what? One apple? Two apples? Half the orchard? Nonsimultaneous arrival makes it hard to see your original grab as treatment at all, unequal or otherwise, thus blocking any easy move from a premise that there are unequal shares to a conclusion that there has been unequal treatment.

What if you grew your orchard from only one apple? Suppose you left the second apple for Ackerman, but Ackerman arrived too late to make use of it. It is not Ackerman's fault that he was late, we may assume for argument's sake, but neither is it your fault. Would that affect what you now owe Ackerman? Why? Did you owe it to Ackerman to turn that second apple into a second orchard, to be claimed by Ackerman whenever he happens to show up?

In Ackerman's original garden, we would feel offended if you grab both apples. Why is the real world so different – so different that if Ackerman were to walk into the cafeteria and say, "Shouldn't I get one of those apples?" we would feel offended by Ackerman's behavior, not yours? Needless to say, the real-world Ackerman would never do that. (He is, in a word, civilized.) So, evidently, there is some difficulty in generalizing from Ackerman's thought experiment. Why? Roughly, the problem is: In our world we do not begin life by dividing a sack of apples that somehow, on its own, made its way to the bargaining table. We start with goods some people have helped to produce and others have not, already possessed and in use by some people as others arrive on the scene. Contractarian thought experiments depict everyone as getting to the table at the same time; it is of central moral importance that the world is not like that. (See Chapter 24 for further discussion.)

Ackerman's short dialogue does an excellent, indeed inspiring, job of capturing the essence of egalitarian concern. In my estimation, the essence and the power of such concern is grounded in an ideal of equal treatment. I do not question that ideal here. The only thing I would say is that the ideal of equal treatment is not the same thing as the idea that we ought to have equal shares.

IF EQUALITY IS FAIR, WHAT ABOUT EQUALIZATION?

David Miller notices a difference between saying equality is good and saying equality is required by justice.[5] If our primary school organizes a track meet, and one boy wins every race and takes every prize, we accept that justice was done. Prizes were fairly won. Still, we are disappointed. It would have been a better (at least more enjoyable) day if the prizes had been spread around. Yet, Miller observes, we need not dress up our disappointment. Not everything that matters is a matter of justice.

We may think a bit more equality would make the world a better place, but we need not insist that each boy claiming a prize *as a matter of justice* would make the world a better (or fairer) place.

DISCUSSION

1. Are we treating people equally or unequally if we try to balance the unequal treatment they receive elsewhere?[6]
2. Recall the scene with which Part 4 began. What counts as giving six-year-old Cindy an equal share: giving her a bicycle now, or on her

[5] David Miller 1999a, 48.

[6] Iris Marion Young holds that we are treating people unequally, and rightly so. She supports "affirmative action programs not on grounds of compensation for past discrimination, but as important means for undermining oppression." She adds, "[A]ffirmative action programs challenge principles of liberal equality more directly than many proponents are willing to admit" and "Supporters of affirmative action policies would be less on the defensive, I suggest, if they positively acknowledged that these policies discriminate, instead of trying to argue that they are an extension of or compatible with a principle of nondiscrimination" (Young 1990, at 12, 192, 195, respectively).

eighth birthday? Can an egalitarian tell Cindy she has to wait? (Of course, if we give Cindy a bike now, her brother will be furious about having to wait two years longer than she did. As noted, equality along one dimension is inequality along another.)[7]

In general, what matters from an egalitarian perspective: that we all get our turn? That we get our turn *at the same time*? What would the latter mean for people born at different times: getting their turn on the same *day*, or at the same *age*?

[7] For an argument that everyone should get their turn at the same time, see McKerlie 1989.

20

What Is Equality for?

THESIS: One can be egalitarian without being humanitarian, but historically the two went together in the liberal tradition, which is why liberal egalitarianism had a point.

EQUALITY AND HUMANITY

Humanitarianism is, roughly, a view that we should care for those who suffer, not only or even mainly as a way of making us more equal but simply because suffering is bad. Humanitarianism concerns how people fare, whereas egalitarianism concerns how people fare *relative to each other.* As Larry Temkin describes them, humanitarians "favor equality *solely* as a means to helping the worse off, and given the choice between redistribution from the better off to the worse off, and identical gains for the worse off with equal, or even greater, gains for the better off, they would see no reason to favor the former over the latter. . . . But such people are not egalitarians in my sense. . . ."[8] True egalitarians want to equalize us even when no one would be better off.[9] Temkin does not mean this as a criticism; he *endorses* egalitarianism in this form.[10] Temkin's objection to humanitarianism is that it is unconcerned with equality per se.[11] "As a plausible analysis of what the egalitarian really cares about, . . . humanitarianism is a nonstarter."[12]

[8] Temkin 1993, 8.
[9] Temkin 1993, 248.
[10] Temkin 1993, 249.
[11] Temkin 1993, 246.
[12] Temkin 1993, 247.

Elizabeth Anderson says, "Those on the left have no less reason than conservatives and libertarians to be disturbed by recent trends in academic egalitarian thought."[13] Academic egalitarians, she thinks, have lost sight of why equality matters.[14] To Anderson, academic egalitarianism gains undeserved credibility when we assume anything calling itself egalitarian must also be humanitarian, but the connection is not automatic.[15] She says, "Recent egalitarian writing has come to be dominated by the view that the fundamental aim of equality is to compensate people for undeserved bad luck." Anderson, though, thinks, "The proper negative aim of egalitarian justice is not to eliminate the impact of brute luck from human affairs, but to end oppression"[16] so that we may "live together in a democratic community, as opposed to a hierarchical one."[17]

If your farm is destroyed by a tornado, you have less, but not because you have been treated unequally. You have not been *treated* at all. If instead the king confiscates your farm because he dislikes your skin color, you have less and also have been treated unequally. In each case, we may want to ease your suffering, but it is only the unequal treatment of the latter case that we have reason to *protest*. Moreover, Anderson suggests, when redistribution's purpose is to make up for bad luck, including the misfortune of being less capable than others, the result in practice is disrespect. "People lay claim to the resources of egalitarian redistribution in virtue of their inferiority to others, not in virtue of their equality to others."[18]

Political equality has no such consequence. In the nineteenth century, when women began to present themselves as having a right to vote, they were presenting themselves not as needy inferiors but as autonomous equals, with a right not to equal shares but to equal treatment.

Two conclusions. (1) Egalitarianism cannot afford to define itself by contrast with humanitarianism. No conception of justice can afford that. (2) Political equality is called for even when economic equality is not.

[13] Anderson 1999, 288. Anderson does not mention Temkin by name.

[14] Price 1999 reaches a similar conclusion. See also Carter 2001.

[15] Anderson 1999, 289.

[16] Anderson 1999, 288. I note that Anderson speaks of the proper "negative" aim, leaving open what she would consider egalitarianism's proper positive aim.

[17] Anderson 1999, 313. As Gaus (2000, 143) describes the liberal egalitarian tradition, "the fundamental human equality is the absence of any natural ranking of individuals into those who command and those who obey."

[18] Anderson 1999, 306.

Thus, to the previous chapter's conclusion that a pluralistic theory of justice can make room for equal shares, we can add that a pluralistic theory can make room for a second equality as well, a specifically political ideal of equal treatment.

GLOBAL EQUALITY IN ECONOMICS AND POLITICS

Iris Marion Young calls it a mistake to try to reduce justice to a more specific idea of distributive justice. Her point applies nicely to discussions of equality. Egalitarianism has a history of being, first and foremost, a concern about how we are treated, not about the size of our shares. Anticipating Elizabeth Anderson, Young says, "[I]nstead of focusing on distribution, a conception of justice should begin with the concepts of domination and oppression."[19] Young sees two problems with the "distributive paradigm." First, it leads us to focus on allocating material goods. Second, while the paradigm can be "metaphorically extended to nonmaterial social goods" such as power, opportunity, and self-respect, the paradigm represents such goods as though they were static quantities to be allocated rather than evolving properties of ongoing relationships.[20]

Some egalitarians agree. Michael Walzer, in particular, says the wealth distribution per se is less important than the possibility of differences in wealth translating into differences in political power, and thus into subordination.[21] As people become rich enough, and have everything else money can buy, what if they begin to buy politicians? Sadly, egalitarians are correct to see this as a problem, for the buying and selling of power is a daily event in democracies and dictatorships alike.

Is there a solution? If it were only a country's own citizens who buy and sell political power, we might dream we could solve the problem by making it illegal for citizens to be wealthy enough to influence legislators. Even if we could stop our fellow citizens from getting rich, though, it is not only fellow citizens who buy our legislators. (Kuwaitis want our legislators to spend our tax dollars on – no surprise – Kuwait.) If politicians have power for sale, then making sure no *citizen* (or special interest group of citizens) can afford to buy it would not solve the problem. We would need to make sure no one *in the world* can afford to buy our politicians and use them to oppress.

[19] Young 1990, 3.
[20] Young 1990, 15–16.
[21] Walzer 1983, 17ff. See also Rawls 2001, 138.

Realistically, if power is being bought and sold, then turned against us, the solution is not to make sure no one is rich enough to buy power but instead to learn how to stop politicians from creating and selling it. Leveling economic shares would not address the real problem. If selling X for a dollar is bad, we go after people who sell X, not after people who have a dollar.

SOCIETY IS NOT A RACE; NO ONE NEEDS TO WIN

Society is not a race. In a race, people need to start on an equal footing. Why? Because a race's purpose is to measure relative performance.[22] By contrast, a *society's* purpose is not to measure relative performance but to be a good place to live. To be a good place to live, a society needs to be a place where people do not face arbitrary bias or exclusion. In liberal society at its best, women, men, blacks, whites, and people of all religions have a real chance to live well, as free and responsible individuals. People need a good footing, not an equal footing.

Recent developments in egalitarian scholarship implictly reflect this insight in two ways. First, egalitarians such as Walzer, Young, and Anderson seem to be regrouping under the banner of an egalitarianism that has roots in a nineteenth-century rebellion against oppression, when egalitarianism was a genuinely liberal movement, allied with nineteenth-century utilitarianism in opposing authoritarian aristocracy. Where this first development recalls the civil libertarianism of nineteenth-century classical liberals and of civil rights leaders of the 1960s, a second development in egalitarian scholarship recalls the humanitarian element of those same movements. What I have in mind is that egalitarians like Richard Arneson are reformulating egalitarianism in such a way that it has a point that can be appreciated even by those who do not already subscribe to a radically egalitarian ideology. "The point of equality I would say is to improve people's life prospects, tilting in favor of those who are worse off, and in favor of those who have done as well as could reasonably be expected with the cards that fate has dealt them."[23]

[22] Admittedly, some people treat society as a race, and measure themselves against the Joneses. But this is no reason for others to care whether those people start on an equal footing. Another thought: We compete against alternative providers of similar services, but alternative providers are people we see out of the corner of our eye, as it were. The people with whom we actually spend our time are partners, not competitors.

[23] Arneson 1999. Many writers reply to Nozick implicitly, but Arneson (2003) responds explicitly and constructively.

The goal is to improve prospects, not to equalize them. We have such things as universities today. To attend them, one no longer needs to be male, or white, or rich. That is progress, not because opportunities are more equal, but because opportunities are *better.*

Pundit Bob Herbert says, "Put the myth of the American Dream aside. The bottom line is that it's becoming increasingly difficult for working Americans to move up in class." More colorfully, "It's like chasing a speedboat with a rowboat."[24] Herbert assumes the American Dream amounts to the suburban cliché "keeping up with the Joneses." I doubt it. I doubt prospective immigrants are daunted by the challenge of running down all the speedboats and thus prefer to emigrate to countries where even the richest are no better than rowboats.

Critics of capitalism once scoffed at the cliché suburban goal of "keeping up with the Joneses." Critics now treat evidence that some group is failing to get ahead of the Joneses as a basis for deeming capitalism a failure. That is what pundits like Herbert and David Cay Johnston are saying when they lament that some group's income *share* has not increased.[25] (Income *shares* add up to 100%; no share can increase unless another share decreases. If I sought to conceal the extent to which society is mutually advantageous, I would equate "getting ahead" not with increasing one's *income* but with increasing one's income *share.*) This change in the dialectic is embarrassing. The old critique of capitalism was thoughtful. It was right to scoff at the goal of keeping up with the Joneses. Elevating that goal to the status of a principle of justice is mindless.

Egalitarianism cannot survive inspection as a call for forcibly maintaining a static pattern of distribution(of income shares, for example), but that is not what liberal egalitarianism was. Societies whose members do not grow and change and distinguish themselves do not survive; a workable egalitarianism makes room for growth and change.[26] There is room, though, within a genuinely liberal theory, for egalitarianism focused on *improving* (not leveling) general life chances – on removing barriers to

24 Source: Bob Herbert, "The Mobility Myth," *New York Times* op-ed, June 6, 2005.

25 David Cay Johnston, "Richest Are Leaving Even the Rich Far Behind," *New York Times* op-ed, June 5, 2005.

26 The same point applies in the international arena. David Miller reflects that if someone in France were getting far better health care than Miller was getting in England, Miller's first instinct would be to suspect that France has a better system and England ought to consider adopting it, not that resources ought to be transferred from France to England so as to reduce inequality. Why? Because the main point of Miller's egalitarianism is to improve life prospects, not to equalize them. See David Miller 1999b.

people bettering themselves not because barriers entrench inequality but simply because barriers are barriers.[27] Societies that succumb to a temptation to experiment with more dictatorial forms of equality must soon either abandon those experiments or be suffocated by them.

[27] Thomas Christiano 2005 argues for an egalitarianism where equality, if not merely a means to the end of well-being, is at least meant to go hand in hand with it. Accordingly, Christiano repudiates the idea of achieving equality by means of leveling down. He holds that if there is an outcome that is better for everyone than the best attainable level of equality, then that superior outcome is better, even from an egalitarian perspective. Christiano also notes that insofar as productivity is a legitimate concern of egalitarian justice, the view to take is that everyone ought to have equal access to the conditions for being productive. If under these conditions, hyperproductive people learn ways to make everyone better off (than the best attainable equality) without making everyone equally well off, so be it.

21

Equal Pay for Equal Work

THESIS: We do not need to choose between equality and meritocracy. Historically, at least in the liberal tradition, they go together.

MERITOCRACY

Chapter 20 asked what egalitarianism has to do with humanitarianism and with resistance to political oppression. This chapter's complementary aim is to ask, what is the connection between equality and merit?

Very roughly, a regime is meritocratic to the extent that people are judged on the merits of their performance. A meritocracy will satisfy a principle of "equal pay for equal work." Rewards will track performance, at least in the long run. A *pure* meritocracy is hard to imagine, but any regime is likely to have meritocratic elements. A corporation is meritocratic as it ties promotions to performance, and departs from meritocracy as it ties promotions to seniority. Note: No one needs to *intend* that rewards track performance. While a culture of meritocracy is often partially a product of deliberate design, a corporation (or especially, a whole society) can be meritocratic to a given degree without anyone having decided to make it so.

Paying us what our work is worth may seem like a paradigm of equal treatment yet it may lead to unequal pay. Norman Daniels says many "proponents of meritocracy have been so concerned with combating the lesser evil of non-meritocratic job placement that they have left unchallenged the greater evil of highly inegalitarian reward schedules. One suspects that an elitist infatuation for such reward schedules lurks behind their ardor for meritocratic job placement."[28]

I suppose this is what Anderson was calling academic egalitarianism. I admit, the kind of view I was taught to consider egalitarian leaves little room to see meritocracy as anything other than an "elitist infatuation." Outside the academy, though, liberal egalitarianism has an older, populist tradition that deployed the concept of meritocracy against hereditary aristocracy.[29] Even the socialist tradition once was partly a meritocratic reaction to a social hierarchy that prevented workers from earning the wages they deserve. Contra Daniels, meritocratic liberalism fought *against* elitism, not for it.[30] Liberalism won. Indeed, liberalism won so decisively that today we hardly remember that a battle had to be fought. In the western world today, no one expects us to bow. No matter how rich or poor we are, the proper way for us to introduce ourselves is with a handshake, which implies that we are meeting as equals.[31] Mundane though that fact is, the very fact that it is mundane – that we can take it for granted – is inspiring.

EQUAL WORTH?

Suppose we have a certain moral worth, and there is nothing we can do to make ourselves more worthy, or less . In this case, we might turn out to be of equal worth. Now suppose instead that, along some dimensions, our moral worth can be affected by our choices. In that case, realistically, there never will be an instant when we are all of equal worth along those dimensions.

What is the true point of the liberal ideal of political equality? Surely not to *stop* us from becoming more worthy along dimensions where our worth can be affected by our choices, but to *facilitate* our becoming more worthy.

Liberal political equality is not premised on the absurd hope that, under ideal conditions, we all turn out to be equally worthy. It presupposes

[28] Daniels 1978, 222.

[29] Galston 1980, 176.

[30] Liberalism also was substantially a revolt against the religious hegemony of the pre-Reformation Catholic Church. Rawls (1996, 303) writes that "A crucial assumption of liberalism is that equal citizens have different and indeed incommensurable and irreconcilable conceptions of the good." Perhaps this is a crucial assumption of recent academic theorizing about liberalism. I doubt it has much to do with liberalism in historical practice.

[31] See Walzer 1983, 249–59.

only a classically liberal optimism regarding the kind of society that results from giving people (all people, so far as this is realistically feasible) a chance to choose worthy ways of life. We do not see people's different contributions as equally valuable, but that was never the point of equal opportunity, and never could be. Why not? Because we do not see even *our own* contributions as equally worthy, let alone everyone's. We're not indifferent to whether we achieve more rather than less. Some of our efforts have excellent results, some do not, and we care about the difference. In everyday life, genuine respect to some extent tracks how we distinguish ourselves as we develop our differing potentials in different ways.[32]

Traditional liberals wanted people – all people – to be as free as possible to pursue their dreams. Accordingly, the equal opportunity of liberal tradition put the emphasis on improving opportunities, not equalizing them.[33] The ideal of "equal pay for equal work," within the tradition from which that ideal emerged, had more in common with meritocracy, and with the equal respect embodied by the concept of meritocracy, than with equal shares per se.

There has been much debate within the academy over what should be equalized. There are hardly any vocal meritocrats in the academy, but if meritocrats were to come forward, they would find they disagreed among themselves in the same way egalitarians do. After all, what are meritocratic rewards supposed to track? Like equality, merit has numerous dimensions: how long people work, how hard people work, how skillfully people work, how much training people need to do the work, how much people are contributing to society, and so on.

32 Temkin makes room for merit: "I think deserved inequalities are not bad *at all*. Rather, what is objectionable is some being worse off than others *through no fault of their own*" (1993, 17). Unfortunately, the room Temkin tries to make has an awkward consequence. If Bill has more than me because he does better work, although mine is good, then according to the quotation's first sentence the inequality is deserved, therefore not bad at all; at the same time, I am worse off through no fault of my own, which is objectionable according to the quotation's second sentence.

33 Richard Miller says "people are pervasively victimized by social barriers to advancement in any reasonably efficient capitalist economy. . . . On the other hand, in an advanced industrial setting, some reasonably efficient capitalist system is best for everyone who is constrained by justice" as Miller conceives of it. There is nothing inconsistent about this, although it "depends on facts that would sadden most observers of the modern industrial scene, saddening different observers for different reasons: Central planning does not work, yet traditional socialists were right in most of their charges of capitalist inequality" (1992, 38).

MERITOCRACY AND MARKET SOCIETY

Markets as Talent Seekers

Meritocracy is not a synonym for market society. Meritocrats often say the market's meritocratic tendencies are too weak; genius too often goes unrecognized. Egalitarians often say such tendencies are too strong; Daniels seems to worry that rewards for satisfying millions of customers are too large. (An especially tedious person will be bitter on both accounts.) Underlying both complaints is a basic fact: Markets react to performance only as such performance is brought to market and offered for sale. So long as Emily Dickinson kept her poetry locked in a drawer, the market had no opinion about its merit. No matter how glorious her genius, her product had to be brought to market before customers could respond.[34]

The other thing to note is that when markets create wealth, they create the possibility of leisure. Markets create time and space within which people can afford to compose poetry (and can acquire paper on which to write it), if that is what pleases them, without having to worry about whether poetry is putting dinner on the table. Markets enable people to build reserves of capital to a point where they can afford to make time for themselves. But markets generally do not judge, and do not reward, what people do with the time they reserve for nonmarket activities.

Markets as Imperfect

David Miller says, "A meritocracy is a society in which people get what they deserve."[35] Economies, Miller notes, are not meritocracies in any systematic way, and eliminating government interference with market mechanisms would not make them so.[36] Miller then says merit should not be allowed to govern the distribution of necessities.[37] An often heard sentiment, but what does it mean? Is Miller saying life's necessities should not go to people who deserve them? Presumably not.[38] An alternative interpretation: Miller is saying macro-level centrally planned distribution of basic necessities should not be done according to merit. This probably is Miller's intent, in which case I agree. Miller acknowledges that, historically, "the arrival of societies in which the market economy has a central role also ushers in desert as a key criterion for assessing the distribution

[34] See Cowen 1998 and Cowen 2000.
[35] Miller 1999a, 198
[36] Miller 1999a, 193.
[37] Miller 1999a, 200.
[38] Miller 1999a, 127. Note Miller's especially apt use of the feminine pronoun.

of goods. For the first time, perhaps, almost everyone can aspire to a state of affairs in which her merits are recognized and duly rewarded."[39] Often enough, Miller says, market prices are a reasonable measure of how much customers want an item, thus of the item's value to customers, thus of what the producer deserves for contributing that much value to customers' lives.[40]

By contrast, a centrally planned meritocracy is a nightmarish idea. Various forms of merit should be recognized and rewarded, but when planners rule, forms of merit unrecognized by them go unrecognized, period. Emily Dickinson or Thomas Edison may have a new idea, but a planner may disapprove, or may doubt his friends will approve, or may be certain his friends will disapprove if the new idea would make their ideas obsolete. And a planner probably would have such friends. Centrally planned meritocracy would in practice be centrally planned mediocrity.

Thus, to sustain the diversity that Miller considers vital to a viable meritocracy, the evaluation of merit must be radically decentralized. Of course, perfection is not an option. The most satisfyingly meritocratic society we can imagine (where evaluation of merit is decentralized, letting us gravitate toward those who appreciate what we have to offer) would never be completely satisfying. Other people will not appreciate our greatest achievements as much as we do.

CONCLUSION

When Martin Luther King said, "I have a dream that my four children will one day live in a nation where they will not be judged by the color of their skin but by the content of their character," he was dreaming of a world where his children could count on equal treatment, not equal shares. He was dreaming of the kind of equality that is not contrary to meritocracy but is instead meritocracy's foundation.

Chapter 20 concluded that egalitarianism cannot afford to define itself by contrast with humanitarianism; no conception of justice can afford that. Likewise, we may add here, no conception of justice can afford to define itself as a repudiation of meritocracy.

DISCUSSION

1. Which is more important as an ideal: society doing all it can to foster people's ability to meet basic needs or society doing all it can to foster excellence?

[39] Miller 1999a, 199–200.
[40] Miller 1999a, 180–9.

2. Theorists sometimes propose, as a way of redressing sexism, that mothers be paid a wage simply for being mothers. The idea: In a traditional family, Dad goes to work and Mom stays home, but they both work hard. Dad gets paid. Why not Mom? (If we abandon a labor theory of value and see payment as flowing not from labor but from customers, then the question becomes: If Dad gets paid only if his customers want what he is selling, why not Mom?) There are many issues. Are domestic wages an issue of gender equality? Equality between sellers who have customers and sellers who don't? Between those who commodify their child-rearing skills (for example, by selling daycare services) and those who don't? How would domestic wages be financed? Would bureaucrats simply be shuffling money between (or within) women's families?

   Especially troubling from an egalitarian perspective: Would upper-class mothers be paid more than working-class mothers? If not, then what would the wage be? Would the wage be set too low to be meaningful to upper-class mothers, or so high that it dwarfs the wages of working-class fathers? If we pay additional wages for each child, and if the wage were meaningful to working class families, thereby leading some working class men to pressure working class wives and daughters to have more babies than they otherwise would, will this liberate women? What if we are not sure? Should we do it anyway?[41]

[41] On the ramifications of shifting childrearing from the "realm of love" to the "realm of money," see Folbre and Nelson 2000. I thank Ulrike Heuer, Ani Satz, and Elizabeth Willott for helpful discussion.

Willott identifies more general issues: Much of our productive work creates positive externalities, making *many* people better off, not only our paying customers. This is as true of parenting work as of any other work. Full-time parents contribute positive externalities, and receive positive externalities in return, as they both sustain and benefit from a society where other people work. The difference: Full-time parents lack paying customers and therefore contribute and receive only these positive externalities. Or rather, they do have clients of a sort, namely their children, but the extent to which children repay has in recent decades diminished. (1) We no longer expect children to repay by some day caring for elderly parents. Social security systems, for better or worse, now replace the payment parents once received from their children. (2) Children once had to stay home to inherit the land, but society has become so wealthy that the value of land is no longer important enough to give adult children a reason to stay home, so they no longer do. (Perhaps unsurprisingly, fertility rates are plummeting, as documented in Willott 2002.)

22

Equality and Opportunity

THESIS: Statistics can mislead, but the numbers seem to say the United States is a vertically mobile and increasingly wealthy society – not a land of literally equal opportunity, to be sure, but for all that still a land of opportunity.

PROGRESS

Have we made progress toward economic equality? How would we know? We have statistical evidence that (in some countries) even least advantaged groups have rising life expectancies, and rising living standards along dimensions we can measure. If we were highly idealistic, we might say rising living standards are not enough: A child's background should have nothing to do with where the child ends up. More realistically, we might say children should have opportunities to be wealthier than their parents were at a comparable age. Progress, so measured, would not be obscured by the truism that upbringing affects a child's life prospects.

One might assume pertinent statistics are readily available, and easy to interpret. Not so. Newspapers often publish articles on the topic, but such articles often are badly mistaken, and it is not easy to do better. I present pertinent data, with painful awareness of how easy it is to be wrong.[42] The information base is always changing, and must be sampled rather than exhaustively reviewed. Moreover, working with available data is a bit

[42] I thank Dr. William Fairley, President and Consulting Statistician for Analysis & Inference Inc. (a company that does statistical consulting and research) for reading and commenting on this chapter. I remain responsible for errors.

like looking for my keys under the streetlamp not because that's where I dropped them but because that's where the light is better.[43] Available statistics shed light on something, but not necessarily on what matters most.[44]

The statistics discussed here depict the United States circa 2002, a country in recession. In recessions, all groups lose ground. Predictably, gaps between groups close. In fact, since so much of this recession was about crashing stock markets, I would have guessed that income gaps would close substantially, disproportionately cutting down richer people who presumably have been receiving more of their income from stocks. I would have been wrong.[45]

HOUSEHOLD INCOME QUINTILES

For the sake of reference, when we divide households into income quintiles, the income cutoffs as of 2002[46] are as follows:

Lowest quintile:	Zero to $17,916
Second quintile:	17,917 to 33,377
Third quintile:	33,378 to 53,162
Fourth quintile:	53,163 to 84,016
Top quintile:	84,017+

Household income at the 80th percentile is 4.7 times household income at the 20th percentile.[47] Is that bad? The question is less important than whether quality of life at the 20th percentile is bad. How bad is it to be earning $17,916? This is not a simple question. Location makes a difference, though. If "household" refers to a single person in a midwestern town, $17,916 will be a lot of money, whereas a single mother

43 Data on inequalities tied to race and gender is sketchy, but I sympathize with those who say such data would be more interesting, by virtue of indicating ways in which inequality tracks factors transparently beyond personal control.

44 For a classic account of how numbers can mislead, see McCloskey 1985.

45 Or at least, I did not see what I expected when I checked statistics for the top *quintile*. However, in a study I had not seen when I wrote the above, Piketty and Saez (2004) say, "The drop in top income shares from 2000 to 2002, concentrated exclusively among the top 1%, is also remarkable. This later phenomenon is likely due to the stock market crash which reduced dramatically the value of stock-options and hence depressed top reported wages and salaries."

46 Source: U.S. Census Bureau, Current Population Reports, P60–221, Table A-3.

47 Household income also is unequally distributed within the top quintile. For example, income at the 95th percentile is $150,002 (ibid), nearly 1.8 times income at the 80th percentile.

raising two children in Boston on that income might not be able to pay the bills.[48]

LARGER HOUSEHOLDS, LARGER INCOMES

Studies of income distribution typically separate populations into quintiles according to household income. While each household income quintile contains 20% of all households by definition, as of 1997 the bottom quintile held 14.8% of individual persons; the top quintile held 24.3%. An average household in the bottom quintile held 1.9 persons and 0.6 workers. Households in the top quintile held an average of 3.1 persons and 2.1 workers.[49] What does this mean? It means one source of statistical inequality is that some households contain more wage earners. If every worker were earning the same income, there would be substantial inequality in household incomes simply because some households house only one worker.

A related point: We can be misled when studying changes in household income in a society where the number of wage earners per household is falling due to such things as rising divorce rates. As the number of wage earners per household falls, average *household* income can fall even as individual incomes rise.[50] Thus, according to the U.S. Department of Commerce's Bureau of Economic Analysis, median household income rose by 6.3% between 1969 and 1996, which on its face seems consistent with the thesis that the economy was more or less stagnant over that period. However, real median income *per capita* rose by 62.2% during the same period.[51]

How could individual income rise by 62.2% when household income rose by hardly one tenth of that? I first thought there must be some mistake. But imagine a household of ten people each earning one hundred dollars in 1969, for a total of one thousand dollars. In 1996, we find that

[48] Tyler Cowen once remarked to me that it is easy to favor equality in countries with homogeneous populations. As a country becomes geographically or ethnically diverse, or welcomes poor immigrants, the costs of equalizing start to rise. It is one thing to achieve a measure of equality in Sweden, or in Kansas. It is another matter to achieve the same measure of equality over a population as large and as diverse as the United States, or Europe.

[49] Rector and Hederman 1999 12, citing U.S. Census Bureau.

[50] If two people live in a typical student household today versus three a generation ago, that would show up in our statistics as a decline in the bottom quintile's average income. Yet, in this case, household income fell because individuals are more wealthy, not less, which is why students now can afford to split the rent with fewer housemates.

[51] McNeil 1998, Table 1.

household income has risen to $1,063, a change of 6.3%. Could incomes of household *members* at the same time rise from $100 to $162? Yes. If the number of householders at the same time dropped from 10 to 6.56, then income per person rose to $162. I could find no data to verify that household size actually fell this much between 1969 and 1996, so the possibility remains that the Commerce Department got its numbers wrong. However, we just showed that its numbers *might* be correct: Personal income could rise by 62% at the same time as household income is rising only 6%. This suffices to prove that if we look only at changes in household income, we see only part of the picture.[52]

In summary, gaps in household income are to some degree accounted for by differences in household size. Also, *falling* household size can make household income appear more stagnant than it really is. Robert Lerman estimates that half the increase in income inequality observed in the late 1980s and early 1990s was due to the increase in number of single-parent households.[53]

OLDER HOUSEHOLDS, LARGER INCOMES

So, some of the difference between income quintiles is a difference in household size. What about differences in age? Some household heads are in their prime earning years while others are not. Suppose we look

[52] A further thought on the demographics of household size: According to Hinderaker and Johnson 1996, 35.

> The Census Bureau keeps statistics separately for 'families' and 'unrelated individuals.' Census Bureau figures show that between 1980 and 1989, real income for the middle quintile of families increased by 8.3%, while real income for the middle quintile of unrelated individuals increased by 16.3%. The CBO [Congressional Budget Office] manipulated this Census Bureau data by combining 'families' and 'unrelated individuals' into the single category of 'families.' Since demographic trends produced more rapid growth in the number of unrelated individuals in the 1980s, and since families headed by two adults on average have far higher incomes than unrelated individuals, combining these groups into a single category greatly depressed average 'family' incomes. Thus, even though the incomes of middle-quintile families increased by 8.3% and the incomes of middle-quintile individuals increased by 16.3%, middle-quintile 'families' in the CBO's sense saw their total incomes decline by 0.8% over the same period.

If it is unclear how a group's income could drop as its component incomes rise by 8.3 and 16.3%, consider a simple illustration. Let family X's income be $1,000 and individual Y's be $100. Average income is $550. Later, family X's income rises to $1,080 while individual Y's rises to $116. Meanwhile, the extra money enables family X's daughter to leave home and live on her own, so there are now two individuals each earning $116. If we then average the three incomes, we get an average income of $437, an apparent $113 drop in *average* income even though *each* income is in fact rising.

[53] Lerman 1996.

again at household income, this time dividing household heads into five age groups, again as of 2002.[54]

a. $27,828 average income when the household head is under 25
b. $45,330 for ages 25–34
c. $53,521 for ages 35–44
d. $59,021 for ages 45–54
e. $47,203 for ages 55–64

Where earnings had once begun to trail off as workers entered their forties, earnings now keep increasing as workers reach their fifties, then drop as early retirements and premature deaths of primary breadwinners begin to cut into the average. Michael Cox and Richard Alm report that, "In 1951, individuals aged 35 to 44 earned 1.6 times as much as those aged 20 to 24, on average. By 1993, the highest paid age group had shifted to the 45 to 54-year-olds, who earned nearly 3.1 times as much as the 20 to 24-year-olds."[55]

Is the increase in statistical inequality also an increase in what we intuitively count as inequality, or is it simply a general increase in lifetime income? If earnings of people edging into their forties now continue to rise when, a few decades ago, earnings would have begun to fall, that will increase statistical inequality as 45–54-year-olds continue to distance themselves from younger counterparts. But is that bad? Is there anyone for whom that is bad? On reflection, a society that enables 45–54-year-olds to make increasingly valuable contributions is a society that on its face satisfies Rawls's difference principle. Statistically, society is more unequal, yet everyone is better off. Everyone's expected lifetime earnings are higher.

The numbers suggest the top quintile is not a separate caste of aristocrats who now earn even more. Instead, median income (not necessarily anyone's personal history, of course) has this trajectory: A median head's income is in the second quintile when the head is under age 25, rises to the third quintile between the ages of 25 and 34, then rises into the fourth quintile and stays there until retirement.

So, when we read that median income at the eightieth percentile jumped by 55% in real (that is, inflation-adjusted) terms between 1967 and 2002,[56] we should understand that for many currently at the twentieth percentile, that jump means increasing lifetime earnings for them, not just for some separate elite. The fact that 45–54-year-olds do

54 Source: U.S. Census Bureau 2003, Table 3.
55 Cox and Alm (1995, 16), citing U.S. Census Bureau.
56 Source: U.S. Census Bureau 2003, Table A-4.

better today, thereby widening gaps between quintiles, is good news in a general way, not only or even mainly for people currently in that age group.

In summary, gaps in household income are partly accounted for by differences in age. Growing gaps may partly be accounted for by improving opportunities for everyone as people move toward peak earning years. Gary Burtless estimates that the proportion of income inequality due to age inequality is 28% among men and 14% among women.[57]

DURING BOOMS, THE RICH GET RICHER, BUT CLASS MEMBERSHIP IS NOT FIXED

If the top quintile's incomes are rising, this is not the same thing as saying the rich are getting richer. In general, the numbers say it is not only that the rich get richer. There also are more people getting rich. In 1967, only 3.1% of U.S. households were making the equivalent of $100,000 in 2002 dollars. By 2002, the number had risen to 14.1%. For whites, the increase was from 3.3 to 15.0%. For blacks, the increase was from 0.9 to 6.6%.[58] (As elsewhere, these numbers are inflation adjusted.) It appears to be false that there was a small cadre of people who had a lot of money in 1967 and today that same cadre has more, and *that* explains why the top quintile has pulled farther ahead. On the contrary, what seems to be happening is that millions upon millions of people are joining the ranks of the rich. They were not rich when they were younger. Their parents were not rich. But they are rich today.

EVEN WHEN MONEY PER SE DOES NOT TRICKLE DOWN, QUALITY OF LIFE DOES

Growing income gaps can mask a narrowing of gaps in quality of life. Inequality would be increasing in a way that mattered if life expectancies of the poor were dropping while those of the rich were rising. In fact, both have risen. I suspect the gap in life expectancy between rich and poor has narrowed (although it stands to reason that extremely unhealthy people make less money and thus tend toward the bottom quintile, while extremely poor people do not live as long, and that each of these facts will affect the bottom quintile's average life span). I found no data directly bearing on that, but I did find this on whites and blacks: Between 1900

[57] Burtless 1990.
[58] Source: U.S. Census Bureau 2003, Table A-1.

and 2001, life expectancy for whites rose 63%, from 47.6 to 77.7 years. Life expectancy for blacks rose 119%, from 33.0 to 72.2 years.[59]

INCOMES ARE NOT STAGNANT

A small minority of economists have said, and countless newspapers have repeated, that middle-class wages are stagnant at best. The balance of evidence points in the opposite direction. Studies showing that average wages fell by, say, 9% between 1975 and 1997 are based upon a discredited way of correcting for inflation (and also ignore a burgeoning of fringe benefits). When we use currently accepted ways of correcting for inflation, the corrected numbers show average wages rising 35% since 1975.[60] In December of 1996, a panel of five economists, commissioned by the Senate Finance Committee and chaired by Michael Boskin, concluded that the consumer price index overstates inflation by about 1.1% per year (as little as 0.8% or as much as 1.6%).[61] If Boskin's figure of 1.1% is correct, then "instead of the stagnation recorded in official statistics, a lower inflation measure would mean that *real* median family income grew from 1973 to 1995 by 36 percent."[62]

Figures for the period of 1996–99 (not a recession, but it is the most recent data I could find) indicate that of those who experienced poverty during this forty-eight-month period, 51.1% were in poverty for two to four months; 5.7% of those who experienced poverty were in poverty for more than thirty-six months.[63] More recent figures, for the trough of the current recession, presumably would show poverty experiences of longer duration. In the neighborhood of the poverty level, we seem to see considerable income mobility, downward as well as upward. People find jobs, and lose jobs. People also retire, thus making permanent downward

59 Source: National Center for Health Statistics at the Center for Disease Control (http://www.cdc.gov/nchs/fastats/). The changes go beyond declines in infant mortality. "Death is on the decline for babies, adults, and older people alike, with AIDS, homicide, cancer, and heart disease all claiming fewer lives." Source: National Center for Health Statistics at the Centers for Disease Control and Prevention, as reported by the Associated Press, September 16, 2002.

60 Both figures are in Norris 1996. The first number is based on the standard Consumer Price Index at the time. The second was supplied by Leonard Nakamura, an economist at the Federal Reserve Bank of Philadelphia.

61 Source: *The Economist* (December 7, 1996) 25. See also Boskin, et al. 1996.

62 Source: *U.S. News and World Report* (September 8, 1997) 104. Emphasis added.

63 Source: U.S. Census Bureau 2003, Figure 6.

moves. Finally, people move to the United States in huge numbers, and first-generation immigrants tend to be impoverished, if only temporarily.

WE APPEAR TO BE A VERTICALLY MOBILE SOCIETY

Suppose bottom quintile income truly had not risen in a generation. What would that have implied? It would *not* mean a group of people were flipping burgers a generation ago and today those same people are still stuck flipping burgers for the same low wage. Rather, had there been wage stagnation since 1967, it would have meant this: When this year's crop of high school grads flips burgers for a year, they will be paid roughly what their parents were paid back in 1967, when they were just out of high school and doing the same jobs. If twentieth percentile wages had been stagnant, the upshot would have been that low-wage *jobs* pay what they always did, not that *people* who once held low-wage jobs remain stuck in low-wage jobs today. Let me emphasize two aspects of this. First, bottom quintile wages have not been stagnant. In inflation-adjusted terms, median income at the twentieth percentile has risen by 31% between 1967 and 2002.[64] Second, these real gains for the twentieth percentile, while good, are of only passing relevance to many now at the twentieth percentile, flipping burgers but planning to move up.

If you think of all of the Americans you know who are over forty but still working, I would bet that every one of them is wealthier today than when he or she was twenty. Their homes and workplaces will be stocked with appliances they could not have afforded (if such appliances existed) when they were twenty. That would not be true in every country, but it is true here.

The U.S. Treasury Department's Office of Tax Analysis found that of people in the bottom income quintile in 1979, 65% moved up two or more quintiles by 1988.[65] Eighty-six percent jumped at least one. That finding is not unique. Using independent data from the Michigan Panel on Income Dynamics, Cox and Alm tracked a group of people occupying the lowest quintile in 1975, and saw 80.3% move up two or more quintiles by 1991.[66] Ninety-five percent moved up at least one.

These studies tracked individual movements. Would findings as dramatic as these be corroborated by studies tracking households rather than

[64] Source: U.S. Census Bureau 2003, Table A-4.
[65] Hubbard, Nunns, and Randolph 1992.
[66] Cox and Alm 1995, 8 citing Panel Study of Income Dynamics data.

individuals? No. Looking at household rather than individual income, Greg Duncan, Johanne Boisjoly, and Timothy Smeeding[67] estimate that 47% of the bottom quintile circa 1975 was still there in 1991. (Actually, Duncan et al look only at nonimmigrant households, which would affect their results. Immigrant households, I would guess, exhibit upward mobility more like that of individual workers.) Twenty percent moved to the top half of the distribution, six percent to the top quintile.

Apparently, then, there is a difference between individual and household mobility. Why? Imagine a household with two teenagers, circa 1975. Two studies then track the household's subsequent history. One study tracks the household members as individuals, and finds that, sixteen years later, the teenagers' incomes have risen several quintiles. A second study tracking the original household qua household finds that the household lost the summer wages that the now-departed teenagers earned while living at home and attending college. The departed teenagers disappear from the second study, because the new and upwardly mobile households they form did not exist in 1975; the second study tracks only households that existed when the study began. So, given the same data, the longitudinal study of households existing in 1975 paints a picture of modest decline while the longitudinal study of individuals suggests volcanic upward mobility. Which picture is more real?

Another perhaps more important difference between the Cox and Alm and the Duncan et al studies: Both studies tracked groups of young people over periods of years, but Cox and Alm tracked how a group member's income changed relative to the general population, whereas Duncan et al tracked how a group member's income changed relative to the group itself. Twenty-five-year olds tend to move up in a general population's income distribution simply by moving toward their peak earnings years over the course of a sixteen-year study. However, if we compare bottom quintile twenty-five-year olds to a control group that likewise moves toward peak earnings years over the course of the study, then normal progress is controlled for, and will not change anyone's relative position within that group. Only abnormal progress – getting ahead of the Joneses – will show up as a rise in relative position. So, it comes down to a question: what is income mobility? Is income mobility about poor people getting ahead of a slowly rising baseline (the general population) or a rapidly rising baseline (a population aging toward peak earning years)? If our main concern is how people fare, then we do well to consult a study

[67] Duncan et al. 1996. Currently available online.

like Cox and Alm's.[68] If our main concern is how people fare relative to each other, then we do better to consult a study like Duncan et al's.[69] I mention such things not to criticize either study but rather to note how different interpretations of income mobility lead to different numbers. We must interpret, not only to understand the numbers, but also to generate the numbers in the first place. There is no algorithm for making these decisions. We are in the realm of art rather than science.

CHILDREN

Peter Gottschalk and Sheldon Danziger separated children into quintiles according to family income.[70] Their data, Michael Weinstein reports, shows that, "About 6 in 10 of the children in the lowest group – the poorest 20 percent – in the early 1970's were still in the bottom group 10 years later. . . . No conceit about mobility, real or imagined, can excuse that unconscionable fact."[71]

Since Weinstein relies solely on Gottschalk and Danziger, I checked the original study. Gottschalk and Danziger were studying American children aged five years or less as the ten year studies began, so that ten years later the children would still be children.[72] What we have, then, is a cohort of mostly young couples with babies, such that ten years later 40% had moved into higher quintiles. Is 40% bad? Out of context, it looks neither bad nor good. Has any society ever done better?

It turns out at least one society has done better: namely, the United States itself. The figure cited by Weinstein is the figure from the first decade of a two-decade study. Weinstein presents the figure from the 1970s (only 43% moving up) as an indictment of America today, neglecting to mention that the study's corresponding figure from the 1980s was 51%. Although the two figures come from the same table (Table 4) in Gottschalk and Danziger's study, Weinstein evidently felt the more

[68] One change I would make would be to find out how many are bottom quintile because they are students. I would not eliminate them from the study, since they are after all poor, but it is no surprise that in a free and vibrant economy, 95% move up. I would try to treat separately those who are not students (and not retired) but who are in the bottom quintile nonetheless. Do they move up too, or are they so small a part of the bottom quintile that, even if they do not move up, 95% of the group that includes them does? I would want to sort this out.

[69] I thank Greg Duncan for his help in sorting this out.

[70] Gottschalk and Danziger 1999. Currently available on Gottschalk's website.

[71] *New York Times* editorial, Feb. 18, 2000.

[72] Gottschalk and Danziger 1999, 4.

up-to-date number and the apparent upward trend were not worth reporting. Weinstein's editorial was published in the country's most prestigious newspaper.

I myself could have been one of those kids that Gottschalk and Danziger are talking about. I grew up on a farm in Saskatchewan. We sold the farm when I was eleven, and moved to the city. Dad became a janitor and Mom became a cashier in a fabric shop. Even before leaving the farm, we had moved up in absolute terms – we got indoor plumbing when I was about three – but we would still have been bottom quintile. Even after we got a flush toilet, water had to be delivered by truck, and was so expensive we flushed the toilet only once a day (and it served a family of eight). Forty years later, my household income is in the top quintile (which implies a far higher level of absolute wealth than that relative position implied forty years ago). Had I been part of Gottschalk and Danziger's study, though, Weinstein would profess to be outraged by the "unconscionable" fact that, when I was ten, I had not yet made my move.

Returning to the study: As I said, I would predict little evidence of upward mobility in a study ending before subjects reach their middle teens. But let us look. Gottschalk and Danziger say that in the 1980s a child's chance of escaping poverty was better than in the 1970s, but the change was not significant.[73] Gottschalk and Danziger say "only one demographic group (children in two parent families) shows a significant decline in the probability of remaining poor."[74] Within that group, the chance of escaping poverty (by which they mean the bottom quintile) went from 47% in the 1970s to 65% in the 1980s. Oddly, the authors parenthetically acknowledge the massively improved prospects of "children of two-parent families," as if that class were a small anomaly that does not bear on their contention that the probability of escaping poverty did not improve.[75]

Finally, again, recall we are talking about people escaping poverty before leaving the ten-to-fifteen age bracket. If we meant to design an

[73] Gottschalk and Danziger 1999, 9.

[74] Gottschalk and Danziger 1999, 10.

[75] Gottschalk and Danziger use the word "poverty" to refer to the bottom quintile. The poverty rate in the United States has risen in the past two years, from a low of 11.3% in 2000 to a 2002 level of 12.1%. So, even during a recession, the bottom quintile (the bottom 20 percent) is no longer the synonym for poverty that it was in, say, 1929, when the poverty rate was 40%. (Source: Levitan 1990, 5–6.)

Source for the more recent numbers: U.S. Census Bureau 2003, Table 2. The poverty threshold varies with household size (and, less intuitively, with age of the householders). For 2004, the official poverty threshold for a household consisting of two adults under age sixty-five is $12,649. Source: U.S. Census Bureau 2005.

experiment guaranteed to show no evidence of upward mobility, we hardly could do better. Yet, amazingly, Gottschalk and Danziger's numbers seem to say 65% of poor kids in unbroken homes escape poverty before earning their first paycheck.

A LAND OF OPPORTUNITY, BUT NOT EQUAL OPPORTUNITY

The United States appears not to be a caste system or a static aristocracy. Racism and sexism remain painfully real, but are neither as prevalent nor as damaging as they were even a generation ago, let alone a century ago. Moving up is not only possible but normal.

None of this even begins to suggest that family background does not matter. Of course it matters. McMurrer, Condon, and Sawhill say the evidence suggests that,

> the playing field is becoming more level in the United States. Socioeconomic origins today are less important than they used to be. Further, such origins have little or no impact for individuals with a college degree, and the ranks of such individuals continue to increase. This growth in access to higher education represents an important vehicle for expanding opportunity. Still, family background continues to matter. While the playing field may be becoming more level, family factors still significantly shape the economic outcomes of children.[76]

According to Gottschalk and Danziger, bottom quintile children living in *single-parent* families had only a 6.4% chance of moving beyond the second quintile.[77] Of course, it stands to reason that single-mother households are not likely to represent the middle of the income distribution.[78] Interestingly, the result of the ten-year study begun in 1971 is that "black children had a higher chance than white children of escaping poverty if they made the transition from a single-parent family to a 2-parent family by the end of the decade(67.9 versus 42.6 percent)."[79] The second study begun in 1981 finds the probability improving to 87.8% for blacks and 57.6% for whites (their Table 6). On a more discouraging note, as of 1998, the percentage of out-of-wedlock births is 21.9% for

[76] McMurrer, Condon, and Sawhill, 1997. The quotation is from the on-line version's conclusion. See www.Urban.gov.

[77] Gottschalk and Danziger 1999, 8.

[78] Again, though, statistics do not always mean what they appear to mean. If parents live together without being married, only the mother's income is treated as belonging to the child's household. So, many children would escape statistical poverty simply by having their parents get married, even if there were no change in actual income.

[79] Gottschalk and Danziger 1999, 11.

non-Hispanic whites and 69.3% for non-Hispanic blacks.[80] I trust even hard-core egalitarians will agree that what is bad about those numbers is how high they are, not how unequal they are.

WHY I WORRY ABOUT THIS CHAPTER

I have cited many sources, but there are none I fully trust. Honesty is not a switch that good people flip and bad people don't. In this arena, honesty is an achievement, an ongoing hard-fought battle.

At a workshop, a panelist handed out photocopied Census Bureau pages to contradict my claims that (a) age is *the* reason for differences in income, and (b) the lowest quintile's income share has increased. Other panelists, puzzled, noted that I made no such claims. We then observed that his photocopied tables showed the lowest quintile's *income* increasing in real terms, even as its income *share* decreased. The panelist began to apologize for not noticing the inconsistency. Other panelists noted that there is none; as per Rawls's difference principle, rising gaps can accompany rising income at the bottom, and the optimal gap from the least-advantaged perspective is not necessarily small. Had this panelist not presented his claims to a gathering of peers, he would have felt certain his data had refuted me. He later said he hoped he was doing me a favor by showing me what sort of reaction I will get, and not always in workshops where misunderstandings are discussed and corrected. I thanked him for caring, and for helping to make this a better book.

We live in a world of incomplete evidence. Statistics appear conclusive, but in fact are not, and will some day be outdated in ways that matter. Data will be compatible with multiple interpretations, subject to refutation by further data. Thus, there can be no guarantee that my interpretations, or anyone's, are correct. Still, I bridge the disciplines of philosophy and economics by training and by profession, and that combination of analytical tools is unusual enough that it seemed wrong not to risk bringing it to the table when the task of analyzing apparently relevant data seems to require it.

DISCUSSION

1. We know that minute differences in economic growth rates, compounded even over the course of a mere century, add up to gigantic differences in prosperity. So, if we believe in the difference

[80] National Center for Health Statistics, "Births: Final Data for 1998."

principle, and if we believe future generations matter, what should we think of ways of redistributing wealth that reduce economic growth?

2. In a world of overlapping generations, rising income shows up in our data as inequality. Suppose the Smiths and Joneses have the same jobs at the same factory, but each year the Joneses get pay raises by virtue of seniority that the Smiths will not get for another three years. Lifetime income evens out, but at no time are wages equal. Is that a problem? Suppose the gap is not three years but a generation, and the gap shows up not in wages but in life expectancy. Suppose life expectancies increase over a century, the Joneses' by 63%, the Smiths' by 119%. Yet, the Smiths' life expectancies remain 5.5 years lower, roughly what the Joneses' were twenty years ago. If it will take twenty years for the Smiths' life expectancies to rise another 5.5 years, is that a problem? Would it be better or worse if life expectancy for the Joneses also rises during that period, and thus a gap persists?

23

On the Utility of Equal Shares

THESIS: Previous chapters discussed synergies between meritocracy, humanitarianism, and equal treatment. This chapter examines a well-known argument about a synergy between equal shares and utility, grounded in the idea of diminishing marginal utility. The argument does not work.

DIMINISHING MARGINAL UTILITY

Thomas Nagel believes that from an impersonal standpoint, if we were picking principles of just distribution from an impartial perspective, we would have to be in favor of radical egalitarianism.[81] At the same time, Nagel realizes, principles of equality are not the only principles we might adopt if we were to consider matters impartially. In particular, utilitarianism embodies its own brand of impartiality, and not everyone would agree that the imperative to equalize matters more than imperatives to maximize utility or to meet basic needs.

Nagel, however, believes that resolving theoretical tensions between equality and utility is moot. Egalitarianism and utilitarianism diverge in theory. As a practical matter, though, they converge in virtue of the phenomenon of diminishing marginal utility (henceforth DMU). As R. M. Hare puts it, the DMU of wealth and consumption means that approaches toward equality tend to increase total utility.[82] Edwin Baker argues that,

[81] Nagel 1991, 65.
[82] Hare 1982, 27. One author who anticipates my argument is Narveson 1997, 292. See also Narveson 1994, 485.

if wealth has declining marginal utility, then "a partial redistribution of income would maximize the total of individual utilities."[83] Therefore, "at least a limited intervention to increase equality will always be justified under utilitarian principles."[84] Abba Lerner says, "[T]otal satisfaction is maximized by that division of income which equalizes the marginal utilities of the incomes of all the individuals in the society."[85] Lerner infers: "If it is desired to maximize the total satisfaction in a society, the rational procedure is to divide income on an egalitarian basis."[86]

Consider that we have a hierarchy of needs.[87] Food could be our first priority even though satisfactions we pursue only after getting enough to eat are greater than anything we get from food. Therefore, what has first priority and what has highest utility need not coincide. (When I got up this morning, eating breakfast came before writing, on my list of priorities, but at the end of the day, the thing I remember as the day's highlight was the writing, not the breakfast.) Theorists, though, tend to assume such cases are atypical.

Suppose it is rational from a personal standpoint for Jane Poor not to patronize the arts with money she needs for groceries. Does it follow that it also is rational from an *impersonal* standpoint for a community not to patronize the arts with money that could have been spent on groceries? If we put ourselves in Jane Poor's shoes, eating first and patronizing the arts later seems rationally imperative. What if impartiality is more a matter of stepping into *no one's* shoes? In that case, we see that hunger relief is not the only impersonal value; it is unclear that the world would be a better place if, say, resources that went into building the pyramids and the Parthenon had instead gone into soup kitchens.

Again, though, most philosophers assume equality and efficiency go hand in hand, and that from an impartial perspective this is a reason to favor equality. John Broome refers to the argument as "the standard utilitarian argument for equality."[88] Thomas Nagel says,

> Even if impartiality were not in this sense egalitarian in itself, it would be egalitarian in its distributive consequences because of the familiar fact of diminishing marginal utility. Within any person's life, an additional thousand dollars added to fifty thousand will be spent on something less important than an additional

[83] Baker 1974, 45.
[84] Baker 1974, 47.
[85] Lerner 1970, 28.
[86] Lerner 1970, 32.
[87] Chapter 26 briefly discusses the seminal work of the psychologist Abraham Maslow.
[88] Broome 1991, 176.

thousand added to five hundred – since we satisfy more important needs before less important ones. And people are similar enough in their basic needs and desires so that something roughly comparable holds between one person and another.[89]

Nagel says we satisfy more important needs before less important ones. Not quite. We satisfy more *urgent* needs first, but the most urgent need is not necessarily most important. What Nagel calls utility is more closely related to urgency than to importance. Utility in this sense is short sighted, a question of what to do with the next available dollar, not a question of what is most worth doing in the grand scheme of things. Nevertheless, DMU is, as Nagel says, a familiar fact. We have all seen cases of one person turning to another and saying, "Here. You need this more than I do." We can all imagine contexts where such words seem not only intelligible but true.

This does not mean, however, that we should join Nagel and others in thinking that DMU resolves the apparent tension between equality and efficiency. In fact, this chapter will show, the tension is real. Further, the tension exists not only in spite of DMU, but sometimes *because* of it.

Whether we should take this as a critique of utilitarianism or of egalitarianism is a matter of perspective. The point here is not to refute egalitarianism, or utilitarianism, but to show that DMU does not reconcile them and, under conditions often assumed to secure their reconciliation, can even worsen the tension between them.

PREMISES

Harry Frankfurt believes the DMU argument is unsound, for it is grounded in false premises. As Frankfurt sees it, the DMU argument makes two assumptions: "[T]he utility provided by or derivable from an *n*th dollar is the same for everyone, and it is less than the utility for anyone of dollar $(n - 1)$. . . . it follows that a marginal dollar always brings less utility to a rich person than to one who is less rich. And this entails that total utility must increase when inequality is reduced by giving a dollar to someone poorer than the person from whom it is taken.[90] Frankfurt thinks both premises are false. First, it is not true that the utility of money invariably decreases at the margin. Second, individuals are not alike; neither is there any reason to suppose their utility functions are alike. Thus, interpersonal comparisons of utility or satisfaction are

89 Nagel 1991, 65.
90 Frankfurt 1987, 25.

problematic. Different people get differing satisfaction from wealth, such that a marginal dollar could be more satisfying to a rich person than to a poor person. We could add that, third, even if the argument were sound, the bureaucracies we set up to undertake egalitarian redistribution tend to be wasteful. Fourth, even if the costs of redistribution are manageable, there can be incentive problems: Redistribution can rob both rich and poor of the incentive to work. From a utilitarian perspective, such costs are at least relevant.

These four responses have some merit, no doubt, but this chapter asks what happens when (1) marginal utilities smoothly diminish, (2) all are known to have the same utility function, so interpersonal comparisons are easy, (3) redistribution is costless, and (4) there are no incentive problems whatsoever. I will show that even in this pristine environment, where the utilitarian case is most straightforward, we have a situation where transferring a dollar from someone who needs it less to someone who needs it more can be unjustified from a strict utilitarian perspective.

Frankfurt says it follows from the premises of the standard utilitarian argument for equality that a marginal dollar always brings less utility to a rich person than to one who is less rich. Let us accept this for argument's sake. This, adds Frankfurt, entails that "total utility must increase when inequality is reduced."[91]

Not so. This chapter explains why not. To see why not, suppose two people, Joe Rich and Jane Poor, have identical and smoothly declining marginal utility functions. For the sake of simplicity, suppose the only good whose distribution is at issue is corn. We take as given, then, that a marginal unit of corn is worth less to a corn-rich person than to a corn-poor person.

Suppose Poor has zero units of corn, whereas Rich has two units. Further, suppose that to have one unit of corn is to have enough to eat, while two units of corn is so much that Rich would get sick if he tried to eat it all. I do not assume that having a unit of corn is a matter of life and death. We may suppose that without corn, Rich and Poor would have to eat something awful, which they could not bring themselves to do if they could eat corn instead. Thus, consuming the first unit has high marginal utility for Rich and Poor alike, while consuming a second unit has low marginal utility. It is easy to see how someone might conclude that total utility increases when we transfer a unit from Rich to Poor, then go on

[91] Frankfurt 1987, 25.

to conclude that the DMU argument for egalitarian redistribution is, at least here, airtight.

THE ARGUMENT

But is it airtight? If it is *possible* that transferring a unit from Rich to Poor in this pristine environment does not maximize utility, then the alleged entailment fails. Notice: We are not trying to prove it *never* maximizes utility to redistribute from people with low marginal utility to people with high marginal utility. To defeat the entailment claim, we need only show it is not *always* maximizing to make such a transfer. The following argument shows just that.

Given one unit of corn, Jane Poor puts it to its highest valued use, namely immediate consumption. Joe Rich, having already consumed a unit and thus being satiated for the moment, invests the corn in something that is, by Rich's own lights, less urgent. Poor eats the corn, whereas Rich, already having eaten enough, has nothing better to do with his surplus than to plant it.

For a person with one unit, consumption is the highest valued use of that unit. For a person with two units, consumption is the highest valued use of the first unit and, because of the diminishing utility of consumption, production is the highest valued use of the second unit. Therefore, if Joe Rich's second unit is transferred to Jane Poor, both units are consumed, whereas if Joe Rich remains in possession of the second unit, then one unit is consumed and one is planted.

In Figure 23.1, C* is the point at which a person with that much corn would rather plant additional corn than eat it. In the story of Rich and Poor, C* equals one unit. Precisely because of diminishing (that is, downward-sloping) marginal utility of consumption, production becomes a higher valued use as wealth (measured on the horizontal axis as units of corn) rises.

Note: Production's tendency to become more desirable relative to consumption is a general consequence of consumption's DMU, and not an artifact of an odd example. The general conclusion: If a community does not have people out that far on their utility curves, so that they have nothing better to do with marginal units of corn than to plant them, the community faces economic stagnation at best.

Therefore, unequivocal utilitarian support for egalitarian redistribution is not to be found in the idea that consumption has DMU. This result in no way depends on questioning the premises of the DMU argument.

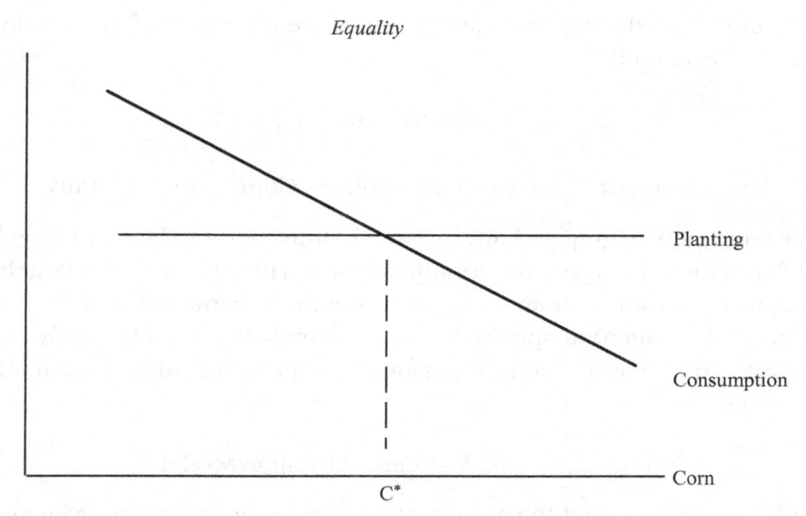

FIGURE 23.1 Marginal Utility of Corn: Planting versus Consuming.[92]

On the contrary, the argument is *grounded* in DMU. Contra Nagel, it does not follow from positing DMU that if everyone counts the same, then a more equal distribution will be better. A society that takes Joe Rich's second unit and gives it to Jane Poor is taking that unit away from someone who, by his own lights, has nothing better to do than plant it and giving it to someone who, by her lights, does have something better to do with it. That sounds good, but in the process, the society takes seed

92 Figure 23.1 is a static snapshot of a dynamic multi-period model in which, if we redistribute in accordance with DMU in period 1, each party consumes one unit by hypothesis. If the utility of consuming a unit equals U, then total utility for period 1 is 2U, and zero thereafter, because no units remain to put into production. By hypothesis, both Rich and Poor are without corn for all subsequent periods, and are left to eat something awful, which they could not bring themselves to do if eating corn were an option.

If we do not redistribute corn in period one, Rich consumes one unit and plants the second, so total utility of corn consumption for period one is 1U. Suppose the productive yield of planted units is $2 + e$ units, so Rich can consume one unit per period and still have more seed corn than he had in the previous period. Eventually Rich's granaries are full, and plowing the surplus back into production has diminishing returns as well, or perhaps he lacks time to do all that plowing by himself. Rich eventually looks for other ways to invest it, such as lending it to Poor, giving it to Poor, or paying Poor to plow for Rich. So the utility of consumption is 1U per period, until we reach a period where accumulating increments of e have added up to a unit of corn or more. At that point, Rich begin to looks for some other way of using the surplus. If he gives or sells a unit of corn per period to Poor, then after that point, the utility of consumption per period is 2 units, indefinitely.

corn out of production and diverts it to current consumption, thereby cannibalizing itself.

RESPONSES

The Argument Assumes a Philosopher's Conception of Utility

The concept of aggregated interpersonal utility used by Hare and Nagel and Frankfurt, which gets the argument going in the first place, has largely disappeared from economic discourse. Kenneth Arrow says that in economics "the utilitarian approach is not currently fashionable, partly for the very good reason that interpersonally comparable utilities are hard to define."[93]

The Argument Assumes a Dynamic Model

With that caveat about the argument's premises, however, Arrow judges the argument to be valid. "In the utilitarian discussion of income distribution, equality of income is derived from the maximization conditions if it is further assumed that individuals have the same utility functions, each with diminishing marginal utility."[94] Arrow is not the only Nobel laureate economist who considers the argument valid. Paul Samuelson reasons that if people are all roughly the same, "so that their utilities can be added, then the dollars gained by the rich do not create so much social welfare or total utility as the dollars lost by the poor."[95] Elsewhere, Samuelson says, "If each extra dollar brings less and less satisfaction to a man, and if the rich and poor are alike in their capacity to enjoy satisfaction, a dollar taxed away from a millionaire and given to a median-income person is supposed to add more to total utility than it subtracts."[96]

Undoubtedly, Arrow and Samuelson would respond by saying they did not mean to suggest the DMU argument would hold in a world of production. They presumably would deny being surprised by the result obtained here, saying they were implicitly if not explicitly assuming the stock of utility-generating goods is fixed.

I accept this response. *If* the DMU argument were treated as relevant only to worlds without production, the argument would be valid, or near enough. Unfortunately, many people, and possibly Arrow and Samuelson too, reason as follows. If the DMU argument's strongly egalitarian

[93] Arrow 1971, 409.
[94] Arrow 1971, 409.
[95] Samuelson 1973, 409.
[96] Samuelson 1973, 423.

conclusions do not quite follow in a world of production, what presumably *does* follow is a suitably weakened version of those same egalitarian conclusions. Not so. In a world of production, DMU can weigh *against* egalitarian redistribution rather than for it, depending on the exact nature of initial endowments and production functions.

The Argument Assumes that Production's Marginal Utility Is Not Decreasing

Figure 23.1 represented the marginal utility of planting as a horizontal line, which we can interpret as constant returns to scale. Under that assumption, there will exist a point C* where DMU weighs against rather than in favor of egalitarian redistribution. (Holding constant other variables, such as production's marginal utility, allows us to focus on how consumption's DMU can lead to production becoming a relatively higher-valued use at C*.) Had we instead assumed increasing returns to scale, representing the marginal product of planting as a rising curve, the same conclusion would follow, for C* would still exist. Suppose we assume decreasing returns to scale, thus representing planting's marginal product as a falling curve. If production's marginal utility is downward sloping, a point C* will exist just in case the production line's slope is gentler than the consumption line's slope (that is, just in case there comes a point when Joe Rich has eaten so much corn that he'd rather bury an additional unit than eat it).[97] And so long as there is a point C*, there is a range where a person is better off planting. If so, then whether consumption's DMU weighs for or against egalitarian redistribution will depend on where we are on the curve, that is, whether we are to the left or to the right of point C*.

In any case, I do not assume production exhibits DMU; neither do I assume otherwise (except for purposes of drawing Figure 23.1). My quarrel is with the idea that *consumption's* DMU necessarily weighs in favor of egalitarian redistribution.

What If We Combine the Utilities of Production and Consumption?

Acknowledging the losses that occur in the course of transferring wealth, Nagel says that, nonetheless, "the rate at which marginal utility diminishes is so rapid that it will still have egalitarian consequences even in many

[97] The point will be to the right of the y-axis if the C-line starts out above the P-line and then falls to meet it. Otherwise, if even the first unit is, improbably, better planted than consumed, C* will be on the y-axis.

cases in which the better off stand to lose more resources than the worse off stand to gain."[98] In Figure 23.1, though, the DMU of consumption for Joe Rich *grounds* the argument against redistribution.

For this argument to work, the marginal utility of consumption for Rich must diminish rapidly enough to dip below the marginal utility of planting by the time Rich allocates his last unit. Otherwise, there will be no point C* at which productive activity becomes relatively attractive. In that case, since Rich's last unit is destined for consumption, it makes utilitarian sense to transfer that unit to someone whose marginal utility of consumption is higher.

So, let me stress: This argument is not against redistribution in general but against assuming that, from a utilitarian perspective, consumption's DMU necessarily weighs in favor of egalitarian redistribution. Neither is this an argument against taxation for the sake of capital investment. Such investments would have to be assessed on their productive merits. Programs aimed at subsidizing the education of poor children could be a wise investment in a society's future. But DMU does not and cannot carry the argumentative weight in such cases.

Redistribution could enhance productivity by putting corn in the hands of people who otherwise would have no chance to become productive, but then we would no longer be redistributing from Rich to Poor; we would be redistributing corn not to poor recipients per se but rather to recipients better positioned to put extra units to productive use. Such recipients may tend to be people who already have C* units of corn, so that they will have the luxury of putting our grants to productive use. Such redistribution could go from the rich to the poor, or from the poor to the middle class. Or, imagine a middle class with more than enough to eat but not enough to invest optimally.[99] Combining the utilities of planting and eating could suggest a case for transferring wealth away from this class in *both* directions – to poor consumers with less to eat, and also (other things equal) to rich producers better positioned to exploit economies of scale.

CONCLUSION

The implications of consumption's DMU are consistently egalitarian only in a model with no production. In a world without production, a

[98] Nagel 1991, 65.

[99] I grew up on an unprofitably small farm (160 acres), so perhaps we were an example of this class.

downward-sloping marginal utility function represents marginal wealth as increasingly frivolous consumption. Utility is maximized in such a world by giving resources to those for whom resources have the most utility. In a world of production, this does not follow. In a world of production, DMU of consumption implies less reason to consume and, relatively speaking, more reason to invest in long-range production. In this world, it is an open question whether utility is maximized by transferring resources to those for whom those resources have the most utility. Utility may instead be maximized by transferring resources to those who will use them in the most productive way.[100]

There is a purpose served by model-simplifying assumptions, but assuming away production possibilities is not like ignoring steps in the utility curve. Ignoring steps simplifies the truth. When we ignore production, though, we do not merely simplify; we ignore *the* prerequisite for meeting needs in the real world.

[100] Needless to say, perverse incentive effects of paying people to *look* needy also afflict institutions that pay people to *look* productive. The latter problem is not unusual in large corporations.

24

The Limits of Equality

THESIS: All principles of justice must answer to social prerequisites of our being able to live good lives together. Among these prerequisites are rules of first possession.

LIBERALISM IS ABOUT FREE ASSOCIATION, NOT ATOMIC ISOLATION

I have been talking about conflicts within liberal egalitarianism, considering how egalitarians can connect equality to meritocratic, humanitarian, and utilitarian considerations. Here I want to step outside that framework to look at the uneasy relationship between egalitarianism and pragmatic considerations underlying customs of ownership by first possession.

In the real world, almost nothing we do is purely distributive. To take from one person and give to another does not only alter a distribution. It also alters the degree to which products are controlled by their producers. To redistribute under real-world conditions, we must alienate producers from their products. The alienation of producers from their products was identified as a problem by Karl Marx, and rightly so; it ought to be regarded as a problem from any perspective.

In a world bound to depart systematically from egalitarian ideals, egalitarian philosophy can encourage these alienated and alienating attitudes. Thus, academic egalitarians, as noted, sometimes see luck as a moral problem, something to resent. On this point, a purist meritocrat agrees, saying success should not be mere luck; it ought to be earned. When meritocratic ideals leave us feeling alienated from a world bound to depart systematically from such ideals, that is regrettable. The point, not directed against egalitarianism in particular, is that even when a radical philosophy is attractive on its face, the psychological baggage that goes with it need not be.

Elizabeth Anderson notes, as many have, that egalitarians "regard the economy as a system of cooperative, joint production" in contrast with "the more familiar image of self-sufficient Robinson Crusoes, producing everything all by themselves until the point of trade" and says we ought to "regard every product of the economy as jointly produced by everyone working together."[101] The Crusoe image is indeed familiar, but only in writings of liberalism's communitarian critics. The liberal ideal is free association, not atomic isolation.[102] Further, the actual history of free association is that we do not become hermits but instead freely organize ourselves into "thick" communities.[103] Hutterites, Mennonites, and other groups moved to North America not because liberal society is where they *can't* form thick communities but because liberal society is where they *can.*

Anderson's point is nonetheless sound. We do not start from scratch. We weave our contribution into an existing tapestry of contributions. We contribute to a system of production, and within limits, are seen as owning our contributions, however humble they may be. That is why people contribute, and that in turn is why we have a system of production.

Obviously, there is much to be said for being thankful that we live within this particular "system of cooperative, joint production" and for respecting what makes it work. When we do reflect on the history of any given ongoing enterprise, we rightly feel grateful to Thomas Edison and all those who actually helped to make the enterprise possible. We could of course resist the urge to feel grateful, insisting that a person's character depends on "fortunate family and social circumstances for which he can claim no credit"[104] and therefore, at least theoretically, there is a form of respect we can have for people even while giving them no credit for the effort and talent they bring to the table. One problem: This sort of

[101] Anderson 1999, 321.

[102] For a seminal expression of the communitarian complaint, see Taylor 1985.

[103] People have been freely organizing themselves into communities of like-minded people since long before there was any such thing as what we now call the state. See Lomasky 2001 and Morris 1998.

[104] Rawls 1971, 104.

"respect" is not the kind that brings producers to the table. It is not the kind that makes communities work.[105]

JUSTICE IN WORLDS WITH HISTORIES

Bruce Ackerman asks us to think about requirements of egalitarian justice by imagining we are on a spaceship looking for new worlds to colonize.

> Coming unexpectedly upon a new world, we scan it from afar and learn that it contains only a single resource, manna... We decide to make this new world our home. As we approach the planet, the spaceship is alive with talk. Since manna is in short supply and universally desired, the question of its initial distribution is on everybody's mind. We instruct the automatic pilot to circle the planet for the time it takes to resolve the question of initial distribution and proceed to the Assembly Hall to discuss the matter further.[106]

Here is the problem: What if the planet is inhabited? Ackerman can stipulate that the world is uninhabited, of course, and in that case we might be comfortable with his conclusions. But what if the planet is inhabited? What if the planet is like, say, Earth?

This is not an idle question. Ackerman wants his thought experiment to illuminate how goods should be redistributed on Earth. So, as we circle over our own planet, dreaming about dividing the stuff, my question remains: What if the planet is inhabited? Are goods on Earth like manna found on an *uninhabited* planet? Or are the real goods Ackerman wants to distribute, here and now, typically goods that have already been discovered, created, possessed, and put to use by someone else?

Philosophers have long sought to identify the essence of justice by asking what people would agree to if they came to a bargaining table to negotiate the outlines of a society in which they subsequently would live. Ackerman is right about this much: If bargainers were looking at a heap of goods on the table, wondering how to divide it, it would not take long for someone to propose that goods be divided into equal shares for all, which would make sense in those circumstances.

What if we change the context just a bit? Suppose people do not arrive at the table simultaneously. Suppose a native tribe got there first.

[105] I am not saying respect for talent and effort are the only forms of respect that make communities work. Elizabeth Willott reminds me of my example from Chapter 8, where a bishop's kindness toward Jean Valjean turns Valjean into a person who deserves the respect implicit in the kindness.

[106] Ackerman 1980, 31.

Now, the ship's passengers and crew are deciding how to divide the tribe's possessions, and a crew member shyly asks, "What makes this our business?"

Or suppose people are born one by one over generations, so that when a new person arrives, there are no unowned goods sitting on a table awaiting fair division. The goods have already been claimed by others, and are at least partly the products of lifetimes of work. We know right away that we will have a harder time knowing what to count as a new arrival's due. We may still conclude, with Ackerman, that justice requires some kind of equal share, but clearly we are in a different situation, requiring a different approach.

Justice would be simpler if the world were as lacking in history as are worlds imagined by contractarians. For better or worse, though, we (and our communities) have histories, and history matters. The point is not that the past must be respected no matter what, but that there are ways of respecting the past that enable people to have mutually respectful, mutually beneficial futures.

FIRST POSSESSION AS AN ALTERNATIVE TO EQUAL SHARES

Chapter 19 discussed Ackerman's view that "equal shares" is a moral default: The rule we automatically go to if we cannot justify anything else. Needless to say, that is not how we actually do it. For various resources in the real world, there is a default position, and it is not equal shares. The practical default is to leave things as they are, respecting the claims of those who got here first.

Carol Rose says rules of first possession (legal rules conferring the status of owner upon the first person unambiguously taking possession of an object) induce discovery. By inducing discovery, such rules induce productive activity. Such rules also help to minimize disputes.[107] They establish presumptive rights that let us claim or concede the right of way, as the case may be, without bloodshed or even wounded pride.[108]

It often seems in philosophical discussion that first possession is confused with exercising raw power when, in fact, first possession functions

[107] Rose 1985.

[108] First possession need not entail permanent ownership. The ownership established by being first to register a *patent* is temporally limited. Or, *usufructuary* ownership lasts only so long as the owned object is used for its customary purpose. Thus, by sitting on a park bench, Bob acquires a right to use it for its customary purpose, but when Bob gets up, the bench reverts to its previous unclaimed status.

in the animal kingdom as one of two *alternatives* to distribution according to raw power. The other alternative to raw power is hierarchies of dominance, through which more powerful males establish a presumption that they will win a fight if it comes to that, thus coming to control resources without actual fighting. First possession is distinct from dominance and from raw power because it secures ownership even for those who are *not* dominant, and even for those who lack the raw power to defend their claims.[109]

Unfortunately, one of our faults as a species is that although respect for first possession is ubiquitous, it tends to be an in-group phenomenon. Individual psychology systematically respects first possession. Group psychology, though, responds to raw power. Groups tend to respect other groups only if those groups can defend their claims in battle. Nothing said here implies that first possession has consistently been respected. On the contrary, aboriginal peoples around the world have been brutally subjugated. Had first possession been respected, many of human history's most tragic episodes would not have happened.

We know this, yet somehow we continue to talk as if justice is about how to *divide* what people contribute, rather than how to *respect* what people contribute.

WHEN EQUALITY IS NOT WHERE WE ARE

Contractarian thought experiments, as noted in Chapter 19, depict everyone as getting to the table at the same time. It is of central moral importance, I said, that the world is not like that. Claims of justice must be fit for the world in which such claims purport to belong. In our world, this means acknowledging that, when any bargainer arrives on the scene, much of the world already is possessed by others in virtue of lifetimes of work (and workers do not find it "arbitrary" that they are the ones who did the work). Theories tend to ignore where we actually are, because theorists want to avoid privileging the status quo, but a theory needs to privilege the status quo in some ways so as to be relevant to it.

Why do property regimes around the world, throughout history, operate consistently according to norms of first possession, not equal shares? The reason, I suppose, starts with the fact that in our world people arrive at different times. When people arrive at different times, equal shares no longer has the intuitive salience it had in the case of simultaneous arrival. When Jane got there first and is peacefully putting her discovery to use,

[109] See Kummer 1991.

then trying to grab a piece of the action, even if only an equal piece, is not a peaceful act.

XENOPHOBIA

An overlooked virtue of first possession: It lets us live together without having to view newcomers as a threat. If we were to regard newcomers as having a claim to an equal share of our holdings, the arrival of newcomers would be inherently threatening. Imagine a town with one hundred people. Each has a one hundred foot wide lot. If someone new shows up, we redraw property lines. Each lot shrinks by one foot, to make room for the new person's equal share (and so on as more people arrive). Question: How friendly will that town be? Even now, in our world, people who see the world in zero-sum terms tend to despise immigrants. They see immigrants as taking jobs rather than as making products, as bidding up rents rather than as stimulating new construction, and so on. The point is not that xenophobia has moral weight, but that xenophobia is real, a variable we want to minimize if we can. Rules of first possession help. What would not help is telling people that newly arriving immigrants have a right to an equal share.

Ackerman believes, "[T]he only liberty worthy of a community of rational persons is a liberty each is ready and willing to justify in conversation."[110] In any viable community, though, most of the structure of daily life literally goes without saying, needing no argument, enabling people to take a "lot" for granted, so they may pour their energy into production rather than into self-defense, verbal or otherwise.

The role played by prior possession in any viable culture, across human history, is an issue for egalitarians, but not only for egalitarians. Every conception of justice needs to make room for it. Meritocracy equally is in a position of having to defer to a norm of respecting prior possession. A viable culture is a web of positive sum games, but a game is positive sum only if players are willing to take what they have as their starting point and carry on from there. A viable conception of justice takes this (along with other prerequisites of positive sum games) as *its* starting point.

LOSING THE RACE

The problem with prior possession, of course, is that those who arrive later do not get an equal share. Is that fair? It depends. How much less,

[110] Ackerman 1983, 63.

exactly, do they get? Some contractarian thought experiments are zero-sum games: First possession leaves latecomers with nothing. For example, in Ackerman's garden, when you grab both apples (or one, for that matter), you leave less for Ackerman or anyone else who comes along later. Thus, as Hillel Steiner has noted,[111] just as first comers would see newcomers as a threat under an equal shares regime, so latecomers would see first comers as a threat under a regime of first possession. Or at least, latecomers would see first comers as a threat if it really were true that, in a first possession regime, it is better to arrive early than late.

But it is not true. One central fact about any developed economy: Latecomers are better off than the first generation of appropriators. We have unprecedented wealth today precisely because our ancestors got here first and began the laborious process of turning society into a vast network of cooperative ventures for mutual advantage. First possessors pay the price of converting resources to productive use. Latecomers reap the benefits.[112] We need to realize that in the race to appropriate, the chance to be a first appropriator is not the prize. The prize is prosperity, and latecomers win big, courtesy of the toil of those who got there first.

So, when someone asks, "Why should first appropriators get to keep the whole value of what they appropriate?" the answer is, they don't. In this world, they keep only a fraction, in the process multiplying rather than subtracting from the stock of what is left for others. It is false that rules of prior possession consign latecomers to a less than equal share. In a society like ours, latecomers are so far faring well, indeed stunningly well compared to those who arrived first.

Latecomers do not fare *equally* well, but grounds for an egalitarian complaint cannot rest on the idea that those with less than their neighbors were made worse off by first appropriators. Latecomers in general (perhaps especially those who would in any case have been in the least advantaged class) are better off, not worse off.

[111] In conversation, September 24, 2000.

[112] See Sanders 2002. It once was thought that the Lockean Proviso – that as much and as good be left for others – has a logic that prohibits original appropriation altogether. (The idea: There are finitely many things in the world, therefore every taking necessarily leaves less for others.) A series of essays (Schmidtz 1990b, Schmidtz 1994, Schmidtz and Willott 2003) observed that appropriating, then regulating access to, scarce resources is precisely how people avert commons tragedies, thereby preserving resources for the future, thereby satisfying the Proviso. When resources are abundant, the Proviso permits appropriation; when resources are scarce, the Proviso *requires* appropriation. People can appropriate without prejudice to future generations. Indeed, when resources are scarce, leaving them in the commons *ruins* future generations.

GROUNDING RULES OF FIRST POSSESSION

Part 4 considered roles that principles of equality properly can play in a theory of justice. To be sure, principles of equality (like other principles of justice) must make room for rules of first possession, but they can. First possession may not itself be a principle of justice, but not every question is a question of what people are due. Sometimes the question is how to resolve disputes over what people are due. Sometimes we resolve disputes by settling who gets to make the call. First possession is to some extent outside the realm of justice, and to some extent corrects justice and keeps justice in its place, for there are times when talking about what people are due is the last thing that would help to resolve conflict. In considering the point of justice, we have to grasp that sometimes what we mainly need to establish is who has the right of way.

We cannot live together without rules that secure our possessions, thereby enabling us to plan our separate lives. When we can count on a general respect for first possession, we need not spend our days in suspense, wondering what we will win and lose in an ongoing war over who will own what. Rules of first possession are signposts by which we navigate in a social world.

Moreover, these signposts are not mere cultural artifacts, any more than is territoriality itself. As virtually any animal capable of locomotion understands at some level, the rule of first possession is the rule of "live and let live." First possession is the rule by which all animals, including humans, understand what to count as an affront and what to count as minding their own business. We can question first possession in theory (and in theory it is easier to attack than to defend), but we do not and cannot question it in our daily practice. We would be lost without it.

PART 5

MEDITATIONS ON NEED

25

Need

"I'm done, Dad!"
"You cut the whole lawn already, Billy? That's great. You did a good job too. So, how much did I say I'd pay you?"
"Five bucks, Dad."
Awkward silence, then Billy softly repeated, "You said five bucks."
"I guess I did, didn't I? You know, I started thinking, and I realized I need the money more than you do. Sorry about that."
"Da-ad . . ."
"Billy, don't look at me like that. I'm only doing what's required by justice."

Should Billy's father give Billy the five dollars? Why? If Billy's father should give Billy the money, what does that tell us? That justice is not about need? Not *only* about need? Should we say justice is about need, indeed only about need, but what people need is context sensitive? In this case, for example, we might say Billy's father should pay, for Billy is, after all, very much in need – not of the money, but of being able to trust his father.

Chapter 26 looks at the big picture, at what people need in the broadest sense. Chapter 27 considers when to distribute according to need. Chapter 28 considers what people need, when distribution according to need is not what they need. Chapter 29 reflects on the place of norms of justice in a functional neighborhood.

To avoid raising false hopes, let me emphasize that I discuss need here not because I can improve upon what others have said about the concept but because I think need-claims are among justice's irreducible primary elements, and because I think the concept of need has useful roles to play in explaining the other primary elements.[1] I have no theory about what people need comparable to my theory about what people deserve, and will not pretend otherwise. I will say what I have to say and then stop, hoping what I have to say is better than nothing.

[1] For excellent discussions (not an exhaustive list!), see David Miller 1999a, chap. 10; Griffin 1986; Braybrooke 1987. And of course, there is Maslow 1970.

26

Hierarchies of Need

THESIS: When we ask what makes one society better than another, there is no reason to work with anything less than the most expansive conception of need: the whole spectrum of human flourishing.

NEEDS VERSUS NEED-CLAIMS

Suppose Michelangelo reaches a delicate stage in his sculpting. Turning to his assistant, he says, "I need the small chisel." His assistant replies, "Do you really *need* the small chisel?"

What is the assistant's point? (1) We can imagine the assistant not fathoming Michelangelo's purpose, yet the context makes it obvious: Michelangelo needs the chisel to continue his work. (2) We can imagine the assistant grasping Michelangelo's purpose, but doubting that the small chisel is apt for that purpose. If that were the case, though, the assistant would not have emphasized the word "need." She would have said something like, "Are you sure you need the *small* chisel?" (3) We can imagine Michelangelo's assistant being at a bad juncture in her philosophical training, playing word games, having developed the skill to make fine distinctions between needs and wants, but not yet having the wisdom to know the point of fine distinctions. Or (4) we can, just barely, imagine the assistant thinking Michelangelo is claiming to be *entitled* to a chisel according to a principle of need-based distribution.

The fourth case is a case where we truly need (that is, our purposes require) fine distinctions between needs and (mere) wants. We know roughly what we mean by the word "need." (Often, to say I need X is to say I want it *now.* Only rarely am I saying that without X, I will die.) For better or worse, this meaning is roughly as much meaning as the word has, barring more precise articulation for use in a specific context. There is no naturally sharp line between needs and wants.

This is not a problem. We do not need a *naturally* sharp line. If and when we need a sharp line, an *artificially* sharp line will do. There is no natural line between driving at a safe speed and driving too fast, either. We could draw an artificial line, but there is no point unless we want, for example, to penalize people on the wrong side of the line. As it happens, we do want to do that, so we *fabricate* bright legal lines, picking thirty miles per hour as a residential speed limit. The artificial bright line is meant to track the inherently vague boundaries of safe speed. The same is true of need. When defining boundaries of need-claims, we fabricate an artificially precise bright line. We do this because we need Michelangelo's *claims* to be sharply limited, not because there is anything naturally sharp about the line between needing a chisel and merely wanting one.

A HIERARCHY OF URGENCY

In Abraham Maslow's theory, there is a hierarchy of needs, with physiological needs forming the base of a pyramid and spiritual transcendence forming the apex, with safety, belonging, esteem, and self-actualization forming some of the intermediate levels. Are any of these levels privileged? Not really. They all matter.[2] If Michelangelo is dying of thirst, then of course drinking water will be more urgent than any need to be sculpting. Part of Maslow's point is that context determines which needs are most urgent. But which needs are most *important*? That is another matter, tied more closely to questions of what counts as a life lived well, and less closely to questions of which need is most urgent at any given moment. A society is better when its citizens can meet their rudimentary survival needs. It also is better when its citizens can afford to look beyond the moment to ask what really matters. The bottom of Maslow's pyramid matters partly because the top matters.

Some contexts (such as when we are theorizing about what people can claim by right) require us to contrive a narrow, precise conception

[2] Needs for safety, love, and esteem are the levels between base and apex. In passing, Maslow says, "The philosopher of ethics has much to learn from a close examination of man's motivational life. If our noblest impulses are seen not as checkreins on the horses, but as themselves horses, and if our animal needs are seen to be of the same nature as our highest needs, how can a sharp dichotomy between them be sustained?" (Maslow 1970, 102).

of need, but we take on that task as the need arises.[3] If we ask not about Michelangelo's need-*claims*, but simply how well a basic structure enables Michelangelo to meet his needs, we get more information if we consider the whole pyramid rather than lop it off at some arbitrary height. When wondering how well a society enables Michelangelo to get what he needs, we are free to ask the obvious: needed for what? We can acknowledge that we have different purposes, and that different purposes imply different needs. So long as Michelangelo is not claiming a right that we meet his needs, we can admit that in a genuine, important, obvious sense, Michelangelo needs to sculpt.

OBJECTIVITY

Michelangelo's needing a chisel is objective insofar as need implicitly is a three-place relation between Michelangelo, the chisel, and a purpose to which Michelangelo will put the chisel. When Michelangelo says he needs the chisel for the next step in his sculpting, the statement has a determinate truth-value, on a par with Michelangelo saying he needs orange juice to avoid scurvy.

This measure of objectivity comes at a cost. If we call need a three-place relation between a person, an item needed, and a purpose for which the person needs the item, then implicitly we are saying need per se lacks independent moral weight. Our need for X has as much weight as the purpose Y for which we need X.[4] Maybe this is not a major cost; so far as I know, we do not say X is needed, thus morally weighty, regardless of whether there is anything X is needed *for*.

[3] For example, as Galston (1980, 163) defines the term, "Needs are the means required for the attainment of urgent ends that are widely if not universally desired."

[4] David Miller (1999a, 206) says we cannot always reply to "I need X in order to Y" with "But do you need Y?" As Miller sees it, this shows there is a sense of need that is not instrumental. If a person needs X to avoid being harmed, then X is an intrinsic need. We do not follow up with "But do you need to avoid being harmed?"

27

Need as a Distributive Principle

THESIS: The only time for distributing according to need is when distributing according to need passes the test of self-inspection.

PASSING SELF-INSPECTION

Distributing according to need is not guaranteed to meet needs. So far as I can see, there is exactly one reason to distribute according to need. Here is the reason: Distributing according to need solves the problem.[5] The point of distributing according to need is not to prove our hearts are in the right place, but to meet the need.

The idea that people ought to get what they need stops calling for distribution *according* to need when distribution according to need stops being what people need. Need-based distribution must, under the circumstances, pass self-inspection.

If parents ought to meet their children's needs as well as they reasonably can – if children are due that much – then that may be a case where justice is about distributing according to need. But need-based distribution is not always what justice requires. The "Lawnmower" case (Chapter 25) is on its face a case where need-based distribution fails self-inspection. In "Lawnmower," what Billy *needs* from his father is recognition that the context calls for distribution according to entitlement,

[5] This is not to say need-based distribution must meet all needs. Instead, it must meet needs in terms of which it is being justified. Also, this is not to say passing self-inspection is sufficient for need-based distribution to be just, only that it is necessary. I thank Arthur Applbaum for such caveats.

where the entitlement is more directly grounded in something like reciprocity than in any principle of need. Need-based distribution can fail the test of self-inspection because alternative principles frequently are more conducive to people meeting their needs.

FOSTERING THE ABILITY TO MEET NEED

Suppose the gold medal, by rule, goes to the fastest runner. Then someone suggests changing the rule to give the medal to the runner who most needs it. What would happen? If you want to elicit speed in a sprint, you reward speed. What are we trying to elicit when we reward need? The point: In many contexts, distributing according to need does not result in people getting what they need. It induces people to do what *manifests* need rather than what *meets* need.

If meeting needs matters, then the test of principle P is not whether it decrees that needs shall somehow be met. The test is whether instituting (or endorsing or acting on) principle P fosters the ability to meet needs. In short, if we care about need – if we *really* care – then we want social structures to allow and encourage people to do what works. Societies that effectively meet needs, historically speaking, have always been those that empower and reward exercises of productive capacities by virtue of which people meet needs.

In the long run, large-scale need-based distribution has never been the key to making people in general less needy. Even if meeting needs were all that mattered, we still would not want to detach the awarding of paychecks, for example, from what actually meets needs, namely, productive work. We still would want resources substantially to be distributed according to productivity.[6]

PRIMARY RULES AND RULES OF RECOGNITION

Here is another way of articulating need's dual role. I elsewhere rely on H.L.A. Hart's distinction between primary and secondary rules.[7] In Hart's legal theory, primary rules are what we normally think of as the law. They define our legal rights and obligations. Secondary rules, especially rules

[6] I do not mean to say there is a short step from productivity's role in meeting needs to rewards being a producer's due. Rather, the point is the metaethical (or transcendental) point being made throughout this book. We are not justifying a conclusion in terms of utility; we are justifying a particular conception of what trumps utility.

[7] See Part II of Schmidtz (1995).

of recognition, tell us what the law is. So, among the primary rules in my town is a law saying the speed limit is thirty miles per hour. A secondary rule by which we recognize the speed limit is: Read the signs. Exceeding speed limits is illegal, but there is no further law obliging us to read signs that post speed limits. So long as I stay within the limit, police do not worry about whether I read the signs. In reading the signs, we follow a secondary rule, not a primary rule.

Recognition rules are not kings among rules of conduct. They do not trump rules of conduct. They do not *win* in cases of conflict. For example, "read the signs" may be the rule by which we recognize rules of the road, but if we found ourselves in a situation where obeying a speed limit would prevent us from reading a traffic sign (perhaps a truck will block our view unless we speed up to pass it), that would not even begin to make the speed limit give way. The highway patrol judges our conduct by the rules of the road, and would be unimpressed if we said we violated the rules of the road out of respect for a higher law bidding us to read the signs.

Therefore, the rules of the road, speed limits and such, do not ultimately reduce to, and do not even answer to, an overarching rule saying "read the signs." The point of the recognition rule is simply to give us a reason for thinking of the rules of the road as requiring one thing rather than another – one speed limit rather than another, and so on. If there were no signs to help us discern the rules of the road (if we knew only that there are rules of the road, and that their overarching point is to encourage us to drive safely), then we would be in the kind of situation we often are in regarding principles of justice. There are few explicit signs to indicate which principles of justice apply in which contexts, but we do know that if we want to *justify* our conception of justice, we must go beyond justice. As noted, we need an argument that does not presuppose the conception for which we mean to be arguing.

Need as a rule of recognition can play that role. Rules of conduct (principles of desert, reciprocity, equality, and need) get recognized; rules of recognition do the recognizing. Need as a rule of recognition will not presuppose any of the rules of conduct (including need-based distribution principles) that make up the conception we are assessing. In particular, the conception of need that we use as a rule of recognition does not presuppose the conception of need that we use in formulating a principle of need as a rule of conduct. Need-claims, as built into a rule of *conduct*, will be truncated in some more or less artificial way. Need as a rule of *recognition* will dictate some such truncation. In other words, how and whether a theory incorporates need-based distributive principles will depend on when and whether need-based distribution really is what people need, that is, on when and whether being able to use certain needs as claims really does help people to live well together.

Need in its role as a rule of recognition will itself be somewhat pluralistic, which will be in some ways a problem. If we cannot settle what level of need in a person's hierarchy of needs is most important (if we cannot settle whether the point of living together is to make sure we do not starve or to give us a chance to sculpt) then it will be as if we have more than one map, with none guaranteed to be unerring. Especially if we take a short run view, we may find metatheoretical desiderata coming into conflict; we may find ourselves having to decide between fostering Michelangelos and helping the least advantaged. In the long run, though, what our children may need (more than guaranteed income, more than vitamins, more than vaccinations, indeed more than anything) is to live in a culture that fosters excellence, a culture where high achievers continually invent better ways to meet whatever needs – vitamins, vaccinations, whatever – their fellow citizens may have.

I described two places where conceptions of need can play pivotal roles in theorizing about justice. First, a suitably *narrow* conception of need can inform need-claims that are best seen as among justice's primary elements, not reducible to claims based on desert, reciprocity, or equality. Second, a suitably *broad* conception of need can serve as a rule of recognition, a rule by which we sort out what merits recognition as a genuine principle of justice. A conception of externalities that we need to internalize in order to live well together can likewise serve as a recognition rule – a reason for endorsing one conception of justice rather than another. It might be nice if we had exactly one unique reason for endorsement that gave simple, unequivocal answers to all questions, but we don't, and a good theory does not pretend otherwise.

DISCUSSION

Should a student who needs an "A" to get into medical school get what she needs or what she deserves? Why? Suppose for argument's sake that nothing matters except need. Would this entail that our aspiring brain surgeon should get the grade she needs rather than the grade she deserves? (What is the general lesson here?)

28

Beyond the Numbers[8]

THESIS: People need to know what to expect from each other.

TROLLEY VERSUS HOSPITAL

Is *promoting* value (doing as much good as you can) the only thing that matters, or is *respecting* value a separate ideal? Under what conditions might they conflict?[9] Here are two notorious philosophical thought experiments.

TROLLEY: A trolley is rolling down the track on its way to killing five people. If you switch the trolley to another track on which there is only one person, you will save five and kill one.

Most people say you ought to switch tracks and kill one to save five. Compare this to:

HOSPITAL: Five patients are dying for lack of suitable organ donors. A UPS delivery person walks into the hospital. You know she is a suitable donor for all five patients. If you kidnap her and harvest her organs, you save five and kill one.

People have a different intuition here. Among students (and U.S. Congressional staffers, at whose workshops I sometimes lecture) that I informally poll, almost everyone responds to HOSPITAL by saying you cannot

8 I especially thank Guido Pincione and Jerry Gaus for helpful e-mails on this chapter.

9 The question is a central coundrum of moral philosophy, taking a wide variety of guises. Andrew Jason Cohen (2004) distinguishes between duties not to interface with autonomy and duties to promote autonomy.

kidnap and murder people, period. Not even to save lives. On a trip to Kazakhstan, I presented the cases to an audience of twenty-one professors from nine post-Soviet republics. They said the same thing. Why? Are the cases really so different? In what way?

TROLLEY tells us numbers matter. Although HOSPITAL seems to have TROLLEY's logical structure, it leads us to a different conclusion. Why? The literature discusses several differences, but one difference I have not heard mentioned is this: HOSPITAL tells us that sometimes what matters is being able to trust others to respect us as separate persons. Hospitals cannot exist, and more generally we cannot live well together, unless we can trust each other to acknowledge that we all have lives of our own. HOSPITAL shows that sometimes we get the best result – a community of people living well together – not by aiming at a result so much as by being trustworthy, so that people can plan to deal with us in mutually beneficial ways.

To a cartoon utilitarian thinking about TROLLEY, all that matters is numbers. But in a more realistic institutional context like HOSPITAL, we intuitively grasp a more fundamental point. Namely, if we don't take seriously rights and separate personhood, we won't get justice; in fact, *we won't even get good numbers.*

ACTS VERSUS PRACTICES

A broadly consequentialist theory needs to treat some topics as beyond the reach of utilitarian calculation. Rights can trump (not merely outweigh) utilitarian calculation even from a broadly consequentialist perspective. Why? Because, from a consequentialist perspective, consequences matter and because, as an empirical matter, there is enormous utility in being able to treat certain parameters as settled, as not even permitting case by case utilitarian reasoning.

Unconstrained maximizers, by definition, optimally use any resources to which they have access, including their neighbors' organs. To get good results in the real world, though, we need to be surrounded not by unconstrained maximizers but by people who respect rights, thereby enabling us to have a system of expectations and trust, which allows us together to transform our world into a world with greater *potential* (a world where delivery companies are willing to serve the hospital). When we cannot count on others to treat us as rights-bearers with separate lives, we are living in a world of lesser potential.

John Stuart Mill famously observed that it is better to be a dissatisfied Socrates than to be a satisfied pig.[10] Of course, it is better to hit an optimum than not, other things equal. On the other hand, Mill's insight is that other things are not equal. If our choice is between making the best of a bad situation versus falling short of making the best of a great situation, we may prefer to fall short, and be a dissatisfied Socrates. Mill, wanting his society to operate as high as possible in utility space, considered it more important to live in a world with a higher ceiling than to make sure every action hits the ceiling. Mill was right.[11]

All optimizing is done with respect to a set of constraints and opportunities. Some of our constraints may be brute facts about the external world, but most will be to some extent self-imposed; some will reflect our beliefs about what morality requires. (We have limited time to spend looking for an apartment, limited money to spend on dinner, and there are things we will not do for money.)[12] We may be constrained not to murder – constrained both by choice and by external factors such as the presence of Joe's bodyguard. If other people can count on us not to murder them, new possibilities open up – opportunities people would not otherwise have. In contrast, if people *cannot* rely on us not to murder them, then our murderous act may be as good as possible under the circumstances – it may hit the utility ceiling, but the ceiling itself will be lower than it would have been had murder been ruled out.[13]

[10] Mill 1979, chap. 2.

[11] According to Geoffrey Sayre-McCord (1996), Hume insisted even more strongly that the consequences of particular cases are not what matter as a general rule. According to Sayre-McCord's "Bauhaus" theory, we do not morally endorse traits based on their actual or even expected utility, but on the utility they *would* tend to have under standard conditions. A related condition: our attention is limited to those who normally might interact with a person having the trait. "We confine our view to that narrow circle, in which any person moves, in order to form a judgment of his moral character. When the natural tendency of his passions leads him to be serviceable and useful within his sphere, we approve of his character . . . " (Hume 1978, 602). Finally, Hume's value theory is pluralistic, allowing incommensurability, and nonaggregative, allowing Humeans to agree with Rawls and Nozick that there is a presumption against sacrificing some for the sake of others. Sayre-McCord endorses Hume's theory so construed, as an account of what we do endorse and also as an account of what we have reason to endorse. I set out my moral theory in Schmidtz (1995). What I call the institutional strand of morality is similar to Sayre-McCord's Bauhaus theory. Over the years, Geoff and I have had so many conversations on this topic that I can hardly guess how much of my view I owe to him.

[12] See Schmidtz 1992 or Schmidtz 1995, Chap. 2).

[13] Someone who says a true utilitarian will take all that into account is saying a true utilitarian will be concerned not so much about consequences of acts as about consequences of practices that permit some kinds of acts and not others, thereby enabling citizens to make

When doctors embrace a prohibition against harvesting organs of healthy patients without consent, doctors give up opportunities to optimize – to hit the ceiling – but *patients* gain opportunities to visit doctors safely. They gain a world with a higher ceiling. Such utility comes from doctors refusing even to ask whether murdering a patient would be optimal.

But what if your doctor really could save five patients by murdering one? Would not a rule letting your doctor do it, just this once, be the rule with the best consequences? Compare this to a question asked by Rawls: In baseball, batters get three strikes, but what if there were a case where, just this once, it would be better if a batter had four?[14] Rawls's insight is that this question presumes to treat "three strikes" as a rule of thumb, to be assessed case by case. Rules of thumb are "rules made to be broken." But in baseball, "three strikes" is a rule of practice, not a rule of thumb. If an umpire were to allow a fourth strike in an exceptional circumstance, baseball would not be able to go on as before.

"Rule of thumb" utilitarians may say, and even believe, they respect the rule against murder, yet they treat whether to obey as a question to decide case by case. By contrast, "rule of practice" utilitarians decline even to *ask* about the utility of particular actions in particular cases. Facing a case where violating a rule would have more utility, rule of practice utilitarians say, "Our theory sorts out alternative practices, like three strikes versus four, by asking which has more utility as the kind of practice that even umpires have no right to evaluate case by case. Our theory *forbids* us to consider consequences in a more case-specific way. We need not say why, but if we did, we would say our being forbidden to consider case-specific consequences has better consequences. For one thing, it gives other people the option of rationally trusting us."[15]

What do we think about a HOSPITAL-like case where we are certain no one will ever know what we've done, therefore certain that our action will not undermine trust? Perhaps it does not matter, since I am not speaking of a world where we can be certain it will never occur to UPS

some kinds of plans rather than others. I agree. My point is only that a true utilitarian in that sense is not an act-utilitarian.

[14] See "Two Concepts of Rules," reprinted in Rawls (1999b).

[15] Is it odd that rule-utilitarianism declines to apply its principle directly to act prescriptions? No. Kant's categorical imperative prescribes maxims, not actions. Eudaimonism prescribes virtues, not actions. In moral philosophy historically, it is act- rather than rule-utilitarianism that is out of step, painting an implausibly simplistic picture of moral life.

Inc. to wonder what is happening to all the delivery personnel they keep sending to our hospital. Suffice it to say, real world morality has the shape it does in part because real world uncertainty is what it is.

Some utilitarians find it a mystery why morality would incorporate any constraints beyond a requirement to do whatever maximizes the good.[16] But from an institutional perspective, there is no mystery. Moral institutions constrain the good's pursuit because the good is pursued by individuals. If the good is to be realized, then institutions – legal, political, economic, and cultural institutions – must get the constraints right, so as to put individuals in a position to pursue the good in a manner conducive to the good's production in general.

There are parallels between rational agents and moral institutions in terms of how they operate in the face of real-world complexity. For example, individuals adopt satisficing strategies in pursuit of particular goals. They impose constraints on local goals so as to bring their various goals into better harmony with each other, thereby making life as a whole go as well as possible.[17] Likewise, moral institutions get the best result not so much by aiming at the best result as by imposing constraints on individual pursuits so as to bring individual pursuits into better harmony with each other. Institutions (hospitals, for example) serve the common good by leaving well enough alone – creating opportunities for mutual benefit, then trusting individuals to take advantage of them. That is how (even from a utilitarian perspective) institutions have a moral mandate to serve the common good that does not collapse into a mandate for ordinary moral agents to maximize utility. In effect, there are two sides to the sense in which institutional utility is based on trust. First, people have to be able to trust their society to treat them as rights-bearers. second, society must in turn trust people to use the opportunities they have as rights-bearers within society.

The kind of consequentialism I have in mind asks us not to maximize utility but to respect existing customs and institutional arrangements that truly have utility. A reflective consequentialist morality is not about one versus five. It is not even about costs versus benefits. It is about how we need to live in order to be glad we are neighbors. It's about getting on with our lives in way that complements rather than hinders our neighbors' efforts to get on with their own.

[16] See Kagan 1989, 121–27. Scheffler (1982, 129) expresses similar skepticism, despite departing from utilitarianism in other respects.

[17] See Schmidtz (1992).

THOUGHT EXPERIMENTS

Philosophical thought experiments tend to be more like TROLLEY than like HOSPITAL, yet the world is more like HOSPITAL. This is a problem, for intuitions elicited by cases like TROLLEY are a misleading guide to cases like HOSPITAL, even if the cases are identical in terms of the numbers at stake. TROLLEY abstracts away from what matters in HOSPITAL. When we think about HOSPITAL, we see that, in this world, people do not need uncertainty. They do not need to be surrounded by unconstrained maximizers. They do not need perfect justice, either. They do need to be able to get on with their lives in peace. *They need to know what to expect from each other.*[18] And when they are trying to cooperate, and thus to coordinate rather than compete, they also need *others* to know what to expect from *them.* They need to be predictable. This is a fact, roughly as objective as the fact that people need vitamin C. It does not presuppose a conception of justice. On the contrary, it is a reason to develop a conception of justice.[19]

It is one thing to catalog intuitions about cases. We risk going astray when theorizing about *why* we have the intuitions we do (a risk I am taking here). Yet, most people have different intuitions about TROLLEY and HOSPITAL, and it is not because the numbers differ. The numbers are the same. Something else is going on. HOSPITAL tells us that most of what is good about our living together starts not with optimizing but with our simply being able to trust each other.

AN "UNTUTORED" REACTION

Wherever I go, whether my audience consists of local students, congressional staffers, or post-Soviet professors, when I present the TROLLEY case

[18] When I say we need to know what to expect from each other, I am thinking not of expectations per se, but of expectations that help us to get on with living together as free and responsible individuals. As Arthur Applbaum (Brown workshop, May 8, 2004) notes, with proper warning we can form expectations about progressive taxation, national health service, rent controls, military conscription, slavery, and forced organ donation. Applbaum's point: *Knowing* that your tax rate is a political football is not like knowing your tax rate is fixed and you need not worry about it. Some expectations let us pour our energy into positive sum games; others help us win zero-sum games. Of course, if we *are* in a zero-sum game, it is better to know, but better yet to know we are not.

[19] One of Sayre-McCord's key insights (1996, 280ff, see note 11) is that our intuitions hold up even as we recognize conditions as nonstandard, and the reliability of our being so predictably swayed *itself* passes the "Bauhaus" test. Why? Because the robustness of our intuition makes us easier to understand, helping us know what to expect from each other, thus helping us to live well together.

and ask them whether they would switch tracks, most will say, "There has to be another way!" A philosophy professor's first reaction to this is to say, "Please, stay on topic. I'm trying to illustrate a point here! To see the point, you need to decide what to do when there is no other way." When I said this to my class of post-Soviet professors, though, they spoke briefly among themselves, then two of them quietly said (as others nodded agreement), "Yes, we understand. We have heard this before. All our lives we were told the few must be sacrificed for the sake of many. We were told there is no other way. But what we were told was a lie. There was always another way."

The more I ponder this reaction, the more I realize how right it is. The real world does not stipulate that there is no other way. (Have you, or anyone you know, ever been in a situation as tragic as TROLLEY? Why not? Have you been unusually lucky?) In any case, I now see more wisdom in the untutored insight that there has to be another way than in what TROLLEY originally was meant to illustrate.[20] As Rawls and Nozick (in different ways) say, justice is about respecting the separateness of persons. If we find ourselves seemingly called upon to sacrifice the few for the sake of the many, justice is about finding another way.

[20] Much of the debate explored differences between acts and omissions. Some of the debate concerned the doctrine of double effect: the difference between killing as a means to an end (HOSPITAL) and killing someone as a foreseen but unintended effect of an action that is compelling on other grounds (TROLLEY). See Foot 1967. For an overview on acts and omissions, see Spector 1992. For a classic discussion of HOSPITAL-type cases, see Thomson 1976.

29

What Do We Need?

THESIS: A conception of justice is a conception of what it takes to be a good neighbor.

A DEVELOPED ECONOMY

What do we need so that our children can live well? In the broadest outline, the first thing we need is a developed economy. New generations tend to be better off than preceding ones just in case they come of age in a developed economy. New generations may not realize they are better off – every generation imagines life was simpler in the old days – but they will be better off in fact.

It was suggested at a workshop that Thomas Aquinas needed little material wealth to flourish in the ways that matter, and my "fetish" with living in a developed economy ignores this fact.[21] It is a curious example: Let us agree that Aquinas's life (1225–1274) was short but sweet. Still, Aquinas was among the *most* advantaged in that society, not the least. It is one thing to say the high priest lives well enough in a materially poor society, and another to say the *least advantaged* live well enough.

A PEACEFUL CULTURE

A second thing we need is to come of age in a peaceful culture: focused on where we can go from here, not on who wronged whom (or worse, which

[21] Chapel Hill Workshop on Capitalism and Morality, April 2004. I thank Chris Morris for the following reply.

ancestors wronged which ancestors) in the past. We must acknowledge that two wrongs cannot be counted on to make a right. Why? Because the wronging does not end with the second; if we commit the second, the second leads to a third, and so on. All sides in a blood feud see themselves as merely evening the score. But our histories are filled with scores that cannot be settled. Unless we acknowledge that the point is to move on, not settle the score, we won't move on. We won't settle the score either.

A CULTURE OF PERSONAL RESPONSIBILITY

A third thing we need is to come of age in a culture of personal responsibility. According to my theory, we assess a society not so much by asking whether people get what they deserve as by asking whether people do something to deserve what they get. The latter question invites us to attend less to the metaphysics of accounting and more to what people can do. According to my theory, children have needs that ought to be met. A society needs to take responsibility, or hold parents responsible, for doing what it takes to meet them. Society likewise needs to foster norms of treating adulthood as an achievement that commands respect and marks when a new adult's needs become fundamentally his or her own responsibility. If we encourage our children to see themselves as entitled to live at someone else's expense, how surprising can it be that encouraging such expectations does more harm than good?

RUDIMENTARY BENEVOLENCE

If there is a fourth thing we need, it might be to face life's challenges with rudimentary benevolence, and thus to stop seeing life as a zero-sum game. When some do better than others, even if *everyone* is doubling his or her life expectancy, some people insist on seeing remaining inequalities as proof that someone is winning at someone else's expense. Rawls says, "It seems clear that society should not do the best it can for those initially more advantaged"[22] but the charitable reading here is that Rawls is thinking only of cases where doing the best for the more advantaged would come at the expense of the less advantaged. To think such cases are *normal* is to think in zero-sum terms.

Being a good neighbor is not a sacrifice.

[22] Rawls, 1999a, 88.

WE DON'T NEED PERFECTION

A society's conception of justice is like a human spine – a functional response to an evolving problem. If spinal design were an engineering problem, we would say that from a functional perspective, the human spine is suboptimal. Starting from scratch, we would design it differently. But we do not start from scratch.

Even if we could start over, the problem solved by our redesign would be the problem we face today, not what we will face tomorrow. If a design is optimal here and now, it will *become* suboptimal. If we had engineered spines for the problem our ancestors faced when they lived in trees, our design would be suboptimal today *even if* it had been optimal at the time. Sometimes, the best we can do is to let our society's imperfect norms evolve along with the problems they solve. Indeed, if our solution to whatever problem is bothering us right now cannot evolve as the problem does, then it never really was much of a solution.

In the biological realm, traits that make a population more exquisitely adapted to a unique ecological niche tend to vanish as the niche does. Traits that persist tend to be those that make populations flexible – able to move into new niches as old ones vanish. I suppose the same is true of a society's norms of justice. The point is not to be a perfect solution to problems of here and now, but to let us cope with here and now without fostering expectations that hinder our adapting to new problems and opportunities. (Imagine a law saying every worker has a right not to be laid off, even if his or her job is to manufacture buggy whips that no longer serve a purpose. In a world of constantly changing possibilities, workers must change with the times in order to continue to have a genuine purpose.)[23]

Some parts of justice (the terrain, not the map) will be fixed. For example, justice will always be about what people are due. And because justice will always be about what people are due, it will never be about punishing the innocent. Some parts of justice may not be analytically built into the concept, and may not be immutable, but it will be hard to

[23] Parts 2, 3, and 4 do not focus on justice being a solution that evolves along with the problem, partly because I wrote them first. (This book has been in progress long enough to count as a product of evolution. One problem with books that take so long to finish is that authors learn to see things in ways their younger selves did not anticipate. One turns a page and finds basic terms being used in a different way, because the two pages were written years apart.)

believe a theory that excludes them. So far as I can see, justice always will have something to do with desert, reciprocity, equality, and need.[24]

CONCLUSION

I explored prospects for a theory that does not try to explain everything, and that uses various tools in explaining what it does. Someone may some day devise a grand unified theory of justice that answers all possible questions about justice. I do not have such a theory, and I suspect that the nature of justice as a cluster concept is such as to preclude a grand unified theory.

I argued that justice is more than one thing, and that the elements composing it are in turn more than one thing. I said we could assess proposals to try to make particular institutions operate according to particular conceptions of justice by asking what happens when such conceptions are given a chance. I do not suppose we could fully agree on what we want to happen, or even on what *did* happen when a conception *was* given a chance. We can agree, though, that fostering prosperity rather than destitution is a plus, a sufficiently important plus that we should investigate how our institutions actually function, not merely theorize.

[24] If norms of justice evolve, and if the analogy with biological evolution is any indication, then the variety of roles played by norms of justice could be as rich as the variety of roles played by bodily organs. At the most general level, all organs contribute to the same function, namely replication, but still they play different roles. I owe this thought to Matt Bedke.

PART 6

THE RIGHT TO DISTRIBUTE

30

Intellectual Debts

The agenda for contemporary philosophical works on justice, including this one, was set in the 1970s by John Rawls and Robert Nozick. Nozick said, "Political philosophers now must either work within Rawls's theory or explain why not."[1] There is truth in the compliment, yet when it came to explaining why not, no one did more than Nozick.

Rawls spent the next three decades responding first to Nozick, then to a barrage of criticism from all directions. In part because of this, no short treatment can capture every nuance of Rawls's evolving theory. A recent book on liberalism by Jon Mandle offers an "overview" of Rawls spanning 133 pages, indicating what a thankless and impossible task it is to try to summarize Rawls in a few paragraphs.[2] However, as is obvious from preceding parts of this book, my theorizing, like most theorizing about justice nowadays, is partly a response to Rawls and Nozick, so Part 6's goal is to consider how they helped to shape the tradition within which this book works. Of course, wisdom simply passed from one generation to the next, then repeated, parrotlike, ceases to be wisdom. Thus, I treat Rawls and Nozick as what they were: thinkers with important insights, not gods.[3] Every generation must relearn and reinvent.

Chapter 31 reconstructs the intuitive core of Rawls's theory. As Chapter 32 explains, Nozick realized that philosophers had tacitly assumed principles of justice must be "patterned." In Nozick's alternative historical

[1] Nozick 1974, 183.
[2] Mandle 2000.
[3] I owe this thought to Bill Edmundson.

theory, principles of rectification play a prominent role. Chapter 33 elaborates. Chapter 34 explores the gap between being arbitrary and being unjust. Chapter 35 asks how we might redesign a Rawlsian thought experiment if our objective were to articulate a pure procedural conception of justice.

31

Rawls

THESIS: Rawls's way of doing philosophy – his sense of what counts as an argument – is unlike mine. Yet Rawls moved the discipline forward. He made progress.

AN ALTERNATIVE TO UTILITARIANISM

According to Rawls, we should think of society as a cooperative venture for mutual advantage. Cooperation enables us all to flourish, but we each want a larger share of cooperation's fruits, so cooperation inevitably involves conflict. One way to resolve the conflict is to distribute the fruits to maximize overall utility. Yet this proposal fails to acknowledge that individuals entering into cooperative ventures are separate persons contributing to those ventures in pursuit of their own legitimate hopes and dreams. Failing to respect their separate projects and contributions is unjust.[4] It may be *the* fundamental injustice.

Standard forms of utilitarianism allow – indeed require – sacrificing the few for the sake of the many (or vice-versa, for that matter) when that would increase aggregate utility. Rawls, though, says that when one person's gain comes at another person's expense, we hardly begin to justify tradeoffs merely by making sure winners win more than losers lose. To Rawls, justice is less like the outcome of a utilitarian calculation and more like the outcome of a bargaining process. Rational contractors meet

[4] How separate are persons, exactly? Does Rawls presume persons are self-contained Robinson Crusoes, as communitarians and feminists sometimes say? See Andrew Jason Cohen 1999.

to negotiate an institutional structure to govern their future interactions, understanding that no one is bound to accept less so that others may prosper. They want a system that promises benefits for all – a system that sacrifices no one for the greater good.

Whatever we think of Rawls's overall theory, Rawls's success in articulating a (perhaps *the*) fundamental problem with utilitarian ethics was a real contribution, as was his idea that conceptions of justice must answer to ideals of reciprocity (*mutual* advantage) rather than to a utilitarian imperative to maximize aggregate value.

JUSTICE AS FAIRNESS

Rawls sought to model justice as a kind of fairness. Many sorts of things can be fair. Evaluations can be fair, or not. *Shares* can be fair, or not. To illustrate the intuition behind Rawls's theory, Christopher Wellman imagines you and I must divide a pie. No one owns the pie, so our only question is whether we can agree on what to count as fair shares. Equal shares is intuitively fair, but what is our procedure for cutting the pie? One answer: I cut the pie into two slices. You pick a slice. I get whichever is left. As Wellman observes, you could have taken the other slice; I could have sliced the pie differently. That is what makes the result fair: not that our slices are equal – they may not be – but that neither of us has grounds for complaint.[5] The process is unbiased, which is to say the process is fair.

Notice: If I am a self-interested maximizer, then my aim is to make the smallest share as big as possible, since the smallest share is what you will leave for me. Now, just as I am concluding that the way to do that is to cut the pie into equal shares, you surprise me with the revelation that you are

[5] Wellman 2002, 66. Rawls describes "I cut, you choose" as illustrating "perfect" procedural justice. "The essential thing is that there is an independent standard for deciding which outcome is just and a procedure guaranteed to lead to it. Pretty clearly, perfect procedural justice is rare, if not impossible, in cases of much practical interest" (1971, 85). Rawls contrasts "perfect" to "pure" procedural justice. "A distinctive feature of pure procedural justice is that the procedure for determining the just result *must actually be carried out*; for in these cases there is *no independent criterion* by reference to which a definite outcome can be known to be just" (1971, 86, emphases added). These definitions entail that Rawls's was a theory of *perfect* rather than *pure* procedural justice, yet Rawls says, "The original position is defined in such a way that it is a state of affairs in which any agreements reached are fair. . . . Thus justice as fairness is able to use the idea of pure procedural justice from the beginning. It is clear then, that the original position is a purely hypothetical situation. Nothing resembling it need ever take place. . . ." (1971, 120).

a baker. You have no great interest in baking for my sake, but because I hold the knife, you consent to me cutting myself a slice of any extra pie you bake, in effect consenting to my treating your talent as if it were a common asset, so long as you get a decent reward for your contributions. So, I make an offer. If I have only one fixed pie, I cut it into equal slices, and we each get half. If you start baking, though, and give me twice as much pie to slice, I will settle for a third of the total. In that case, I get two thirds; you get four thirds. We both win. If you agree, we shake hands and launch our society as a cooperative partnership.

To summarize, we initially assume we are entitled to an equal share of the pie, but realize we can make the pie bigger by encouraging each other to work harder. We encourage each other by rewarding efforts to make the pie bigger: offering more pie to those who do more work. In effect, we allow inequalities if and when doing so makes us better off. I call this the precursor.

PRECURSOR: Inequalities are to be arranged to the advantage of everyone affected: Everyone should get more than what *would have been* an equal share in a more equal but less productive scheme.[6]

The precursor leaves us with an unanswered question, though. Looking back at the pie, why did I propose a split of one third/two thirds? I could have offered less. Would that have been wrong? What I call the precursor is "at best an incomplete principle for ordering distributions."[7] It tells us that departures from equal shares are just only if they make everyone better off. However, "there are indefinitely many ways in which all may be advantaged when the initial arrangement of equality is taken

[6] See Rawls 1971, 60, 62. In addition to entering Rawls's argument as the difference principle's logical precursor, what I call the precursor seemingly was temporally prior too, appearing in Rawls's writing as early as 1958. Amazingly, in his first published statement of the two principles, the second principle is not the difference principle! It is closer to what I call the precursor. Here is the passage:

> The conception of justice which I want to develop may be stated in the form of two principles as follows: first, each person participating in a practice, or affected by it, has an equal right to the most extensive liberty compatible with a like liberty for all; and second, inequalities are arbitrary unless it is reasonable to expect that they will work out for everyone's advantage, and provided the positions and offices to which they attach, or from which they may be gained, are open to all.

See Rawls's "Justice as Fairness," (1999b, at 48; 1st pub. 1958). See also "Constitutional Liberty and the Concept of Justice," (1999b, at 76; 1st pub. 1963).

[7] John Rawls, "Distributive Justice," (1999b, at 135, 1st pub. 1967).

as a benchmark. How then are we to choose among these possibilities?"[8] A complete theory specifies how to divide the gains.

Rawls's way of completing the precursor is to target one position and maximize the prospects of persons in that position. Which position should be so favored? Some would pick bakers. Rawls, though, picks the least advantaged – the group to which life has otherwise been least kind. Roughly, we make the smallest share as big as it can be. We thus arrive at the difference principle.

DIFFERENCE PRINCIPLE: Inequalities are to be arranged to the greatest advantage of the least advantaged.

THE LEAST ADVANTAGED CLASS

Who are the least advantaged? First, Rawls says the phrase refers not to a least advantaged person but to a least advantaged economic class. Second, the class is identified by wealth and income, not by any other demographic.[9] In Rawls's theory, "least advantaged" refers in practice to typical representatives of the lowest income class, no more, no less.[10] Their disadvantages and needs are not unusual but are instead stipulated to be "within the normal range."[11]

Why does Rawls use a name like "least advantaged" to refer to people who are not literally the least advantaged? Rawls is trying to walk a fine line. He wants to articulate a sense of justice that not only embodies compassion for the less fortunate, but that also embodies an ideal of reciprocity that distinguishes his theory from utilitarianism. (That is, he not only wants to say utilitarianism attaches too little weight to the least advantaged, but that it attaches weight for the wrong reason.) "The first problem of justice concerns the relations among those who in the everyday course of things are full and active participants in society."[12] For this reason, Rawls imagines a situation where there is no such thing as a class that other classes do not need. "The intuitive idea is that since everyone's well-being depends upon a scheme of cooperation without which no one could have a satisfactory life, the division of advantages should be such as to draw forth the willing cooperation of everyone taking part

[8] Rawls 1971, 65.

[9] This assumes basic rights are secure for all. For example, no class is disadvantaged by virtue of being enslaved. Rawls 2001, 59n.

[10] Rawls ("Distributive Justice," in 1999b, 139), (1971, 98), (2001, 59).

[11] Rawls 1999a, 83.

[12] Rawls 1999a, 84.

in it, including those less well situated."[13] The least advantaged, as Rawls defines them, are least advantaged *workers*, not least advantaged *people*. They claim a share of the social product because they contribute to it, not because they need it. "The least advantaged are not, if all goes well, the unfortunate and unlucky – objects of our charity and compassion, much less our pity – but those to whom reciprocity is owed."[14]

STRAINS OF COMMITMENT

Is the least advantaged the only class that matters? Taken at face value, the difference principle says that (to borrow Rawls's example) if we can make the least advantaged better off in the amount of one penny, we must do so even when the cost to others is a billion dollars. To say the least, this "seems extraordinary."[15] However, Rawls says, "the difference principle is not intended to apply to such abstract possibilities."[16] Whether the principle applies, though, has nothing to do with whether the possibility is abstract. For that matter, whether the principle applies has nothing to do with whether Rawls *intended* it to apply. Suppose Joe says, "There are no two-digit prime numbers." Jane asks, "What about eleven?" Joe responds, "The principle is not intended to apply to such abstract possibilities." Obviously, Joe has to do better. So does Rawls, but better answers are ready to hand.

First, Rawls did not intend to impose excessive "strains of commitment."[17] We must not ask so much as to make compliance unlikely. Rawls mainly has in mind not asking too much of the less advantaged, but it is vital that the system not ask too much of those whose contributions are most vital. Pushing them so hard that they emigrate – because we took a billion, or because only a penny made its way to the less advantaged – would be bad for all, including the less advantaged.

Second, although the principle may apply to many "abstract" possibilities, it does not apply to case-by-case redistribution. The difference

[13] Rawls 1971, 15.

[14] Rawls 2001, 139. See also Stark 2004.

[15] Rawls 1971, 157.

[16] Rawls 1971, 157. Rawls offers two further answers. First, the difference principle is prevented from applying to such questions by lexically prior principles of justice. Second, the problem tends not to arise in practice because "in a competitive economy (with or without private ownership) with an open class system excessive inequalities will not be the rule. . . . Great disparities will not long persist" (1971, 86).

[17] Rawls 2001, 104.

principle applies only to a choice of society's basic structure. Is this restriction of scope ad hoc, as Rawls's critics often say? No! Why not? Because applying the principle to every decision, as if Joe should never earn or spend a dollar unless he can prove that doing so is to the greatest benefit of the least advantaged, would cripple the economy, hurting everyone, including the least advantaged. The difference principle rules out institutions that work to the detriment of the least advantaged, including ones that overzealously *apply the difference principle* to the detriment of the least advantaged. There is nothing ad hoc about this constraint. It derives straightforwardly from the difference principle itself.

THE VEIL OF IGNORANCE

The core principles of Rawls's theory are these:

> 1. Each person is equally entitled to the most extensive sphere of liberty compatible with a like liberty for all.
>
> 2. Social and economic inequalities are to be arranged so that they are (a) to the greatest advantage of the least advantaged, and (b) attached to offices and positions open to all under conditions of fair equality of opportunity.[18]

Rawls says the first principle takes precedence, where precedence "means that liberty can be restricted only for the sake of liberty itself."[19] There has been relatively little discussion of the first principle, and few quarrel with it.[20] The controversy surrounds the second, particularly part (a), known as the difference principle.

It is easy to envision *skilled* workers objecting to the difference principle, saying "You imagine us negotiating terms of cooperation, then conclude that *fairness* is about someone else getting as much as possible? We never asked you to see us as self-made Robinson Crusoes, only that you grant that our growing up in 'your' society does not affect our status

[18] Rawls 1971, 302. Rawls sometimes adds to part (a) a qualification that the difference principle must be consistent with a principle of just savings.

[19] Rawls 1971, 244.

[20] In a little-noticed passage, Rawls retracts the idea that his first principle has "lexical" priority: "While it seems clear that, in general, a lexical order cannot be strictly correct, it may be an illuminating approximation under certain special though significant conditions" (1971, 45). Rawls likewise retracts the claim that the first principle calls literally for a "most extensive" system of liberty, indeed going so far as to say, "No priority is assigned to liberty as such, as if the exercise of something called 'liberty' has a preeminent value. . . ." See Rawls 1996, 291. See also Rawls 2001, 42. For a critique of Rawls's retreat from the original maximalist formulation, see Loren Lomasky 2005.

as separate persons, and by itself gives you no claim whatever to any skills we bring to the table."

Rawls, in essence, has two replies. First, he argues, skilled workers are wrong to see themselves as mere means to the ends of the less advantaged. Workers make this mistake if they see their skills as part of them rather than as accidents that befell them. When skilled workers reconceive themselves as undeserving winners in a genetic and social lottery, they will see the distribution of skill as a common asset. They will not see themselves as individually bringing *anything* to the table, other than their interests.

Second, Rawls says, skilled laborers would not complain if they could view the difference principle as what they would choose in a fair bargaining situation. What would be fair? Suppose Jane assesses alternative distributions from behind a *veil of ignorance*, not knowing which position in the distribution she will occupy. If Jane has no idea whether she is a manual worker or a mid-level manager, she will not try to bias the arrangement in favor of managers. She will seek an option that is good for all.[21]

"LEAST ADVANTAGED" IS NOT A RIGID CATEGORY

I said if Jane has no idea whether she is a worker or a manager, she will not try to bias arrangements in favor of managers. She will seek an option that is good for all. So why would Jane pick the difference principle, which seems biased in favor of workers, unless Jane knows she is a worker? The question divides Rawls scholars.[22]

One reason to see the difference principle as less biased toward a given group than it may appear, though, is that "least advantaged" is a fluid category. If Joe starts as least advantaged, then a system that works

[21] In a later essay, Rawls took a different approach, saying the contractarian thought experiment was a dispensable heuristic for envisioning what it would be like (in Kantian terms) to strip away our contingent phenomenal characteristics and choose as pure noumenal selves. Since our noumenal selves are identical – mere tokens of the essence of rationality – negotiation is superfluous, and the veil of ignorance now models how a single noumenal self would choose. This avoids problems with hypothetical bargaining but abandons the contractarian commitment to explicitly building respect for the separateness of persons into the theory's foundation. See John Rawls, "Kantian Constructivism in Moral Theory," (1999b, 1st pub. 1980).

[22] Many readers find it hard to see the difference principle as fair, or as what a rational agent would choose. John Harsanyi (1955), seen by some as having invented the veil of ignorance thought experiment, held that the rational choice from behind a veil would be a principle of utility.

well on his behalf eventually puts him in a better position than Jane, at which point the system turns to doing what it can for Jane, until she is sufficiently well off to make it someone else's turn, and so on.

Problem: Is this fluidity real? Yes, or at least, it would be real in a society satisfying Rawls's difference principle. When Karl Marx was writing in the mid-nineteenth century, Europe divided into rather rigid social classes, defined by birth. When Rawls began writing in the 1950s, Marxism was still influential among intellectuals, and Marxists tended to speak as if things had not changed. *If* we lived in a rigid caste society where sons of manual workers were fated to be manual workers (and their sisters were fated to be wives of manual workers), then their best bet would be to set a minimum wage as high as possible without going so far as to make it unprofitable to hire them.

Now suppose a worker has an alternative: Emigrate to a fluid society where manual work pays less than it might, but where his sons and daughters can go to college and be upwardly mobile. Does he stay or go? Should what poor people want out of life affect our thinking about what is to their advantage?

Here is an interesting feature of the more fluid world: Higher income classes can consist substantially of people who once were (or whose parents were) manual workers themselves. Higher income classes, accordingly, can consist substantially of people advantaged *because* they grew up in a world where people born poor – as they once were – had a chance to move up.

In a vertically mobile society, there will be a big difference between unskilled workers who have what it takes to move up and unskilled workers who, for whatever reason, do not. But note that this is a big difference *only* in vertically mobile societies. Behind the veil of ignorance, we have not yet chosen to create a vertically mobile society; within a Rawlsian framework, these subgroups have similar prospects *until we decide otherwise.* Behind the veil, we *decide* whether talented young people should not be stopped by accidents of gender or class.

This fluidity is not what Rawls had in mind. Rawls was not envisioning a world where working-class Joe can acquire skills that will make him wealthy later in life. But if that fluid world is best for Joe, then Rawls is right: What works for Joe works for every class, because in that world, higher income classes will contain substantial numbers of people who started (or whose parents started) as Joe did and made the most of the opportunity. If income mobility *is* to the benefit of younger people who earn less, then by the same token it *was* to the benefit of older people who once earned low wages themselves, then moved up. (Nozick's critics want to portray him as defending the rich, but in his own mind Nozick, himself born poor, was defending the legitimacy of a poor person's dream of a better life.)

It comes down to a question: Are fairness and security the same thing? If not, is the difference principle supposed to concern fairness or security? Suppose the answer is fairness. In that case, if we ask which basic structure is best for the least advantaged, it may turn out to be the one guaranteeing the highest minimum wage. Or the best system may, without guaranteeing much of anything, offer people the best chance to upgrade their skills and thereby earn more than they would in a system with higher minimum wages but less upward mobility. This is settled more by experience than by theorizing. What theorizing settles, in Rawls's eyes, is that making the less advantaged better off is the result to look for when evaluating societies.

In the end, there also is something to say on behalf of Rawls's insistence that we talk of classes, not individuals. There is a cliché: A rising tide lifts all boats. This cliché, as Rawls knew, is not literally true, but good societies make it roughly true. Rawls saw that even in the system he favors, people would fall through the cracks. The tide will never benefit literally every person, but good institutions can and do make it true that the tide lifts all income *classes.* (Even this is hard to guarantee, though. Rawls is being realistic when he says that even in his favored system, we could only "hope that an underclass will not exist; or, if there is a small such class, that it is the result of social conditions we do not know how to change.")[23] Even the least advantaged class shares in the tide of health benefits (life expectancy, safe water, immunizations) and wealth benefits (electricity, shoes) created by cooperation. If all goes well, that is, if Rawls's difference principle (or his precursor) is satisfied, then whole classes will not be left behind by the tide people create when they contribute their talents to cooperative ventures in a free society.

I have not been treating the difference principle as a principle of distribution. Readers often interpret it that way, but Rawls himself meant to evaluate basic structures in terms of how well the least advantaged actually fare. On the most canonical reading of Rawls, if the least advantaged (systematically) do better in system X than in system Y, then system X is more just by the lights of the difference principle. This is so even if X's

[23] Rawls 2001, 140.

government makes no official promises, while Y's government officially promises the moon.

We can imagine Rawls saying Y is just only if its government officially dedicates itself to satisfying Rawls's principles.[24] But this would be refuted by the same argument by which Rawls refutes strict egalitarianism: Namely, if X were better for all, rational agents would choose it over Y regardless of whether system X is less equal, makes fewer promises, or has a different declared aim.

MUST THE PRINCIPLE OF MAXIMUM EQUAL LIBERTY TAKE PRIORITY?

Many who reject Rawls do so on the grounds that (they think) Rawls's theory makes too little room for the primacy of liberty. Those who embrace Rawls tend likewise to brush past his first principle. Here is one reason, though, for taking seriously the idea that Rawls's first principle really does come first, and liberties are not to be sacrificed even for the sake of the least advantaged. Suppose we found that Jim Crow laws worked to the greatest advantage of the least advantaged.[25] Such a result would make the difference principle hard to believe, at least if our theory looked to the difference principle to answer such questions. But what if our theory looks instead to a principle of maximum equal liberty? On that theory – that is, Rawls's theory – the first principle rules out Jim Crow laws from the start. On Rawls's theory, it *doesn't matter* whether such laws satisfy the difference principle. And *that* seems like the right result.

In Rawls's actual theory, liberty is primary. Moreover, making the difference principle primary would hurt the least advantaged. It is *not* in their interest that their liberty be in the hands of the politically advantaged – the sort of legislators who pass Jim Crow laws. The least advantaged must be able to count on their liberty not being a political football, not being what legislators can sacrifice in pursuit of paternalistic goals (for

[24] Rawls comes close to saying this in 2001, 137, 162. See also Brennan 2004.

[25] "Jim Crow" refers to a body of law in the southern states aimed at segregating races. So, for example, section 369 of Birmingham's Racial Segregation Ordinances states: "It shall be unlawful to conduct a restaurant or other place for the serving of food in the city, at which white and colored people are served in the same room, unless such white and colored persons are effectually separated by a solid partition extending from the floor upward to a distance of seven feet or higher, and unless a separate entrance from the street is provided for each compartment." (Source: Courtesy of the Birmingham Civil Rights Institute.)

example, prescription drug benefits) that in practice always look so much more urgent and concrete than ideals of liberty.[26]

In Rawls's theory, the difference principle does not dictate how much liberty we should have. It is the other way around: A commitment to maximum equal liberty dictates the scope legislators have to arrange basic structure for the greatest benefit of the least advantaged. In a political world where the best laid plans go awry more often than not, this is as it should be, especially from the perspective of the most vulnerable. Again, however critical one may be of Rawls's overall theory, this is a real achievement. Or a constellation of achievements: Insights that (a) society at its best is a cooperative venture for mutual advantage, (b) no class of boats should be left behind by the rising tide of wealth in a modern economy, not even the least advantaged class, and (c) liberty is a precondition of there being a rising tide in the first place, especially the kind of tide that lifts *all* classes.

ARTICULATING A CONSENSUS

Before examining Nozick's response to Rawls, we should repeat that Rawls's closest followers disagree, sometimes ferociously, over how Rawls is best interpreted and defended. There are hundreds of theories about how to pull together Rawls's arguments and about which to discard so that the remainder can be an internally consistent whole. In later years, Rawls came to view his work not as a proof that his two principles are true, but instead as a way of articulating beliefs he considered implicit in contemporary Western democracies.[27]

Many of Rawls's followers were distressed by this seemingly colossal retreat. Yet, Rawls's later interpretation was exactly right. Rawls gave us a vision, a vision with grandeur, even if it could not withstand scrutiny as the deductive proof that some people wanted it to be. Rawls was saying that, despite differences in our comprehensive moral views, there is an overlapping consensus implicit in how we live together in Western democracies.

Accordingly, our task is not to dwell on details, but to reflect on the extent to which we share this grand vision: (a) Liberty comes first, and

[26] Rawls says, "[I]n a just society the liberties of equal citizenship are taken as settled; the rights secured by justice are not subject to political bargaining" (1971, 4).

[27] This view is developed in "Justice as Fairness: Political, Not Metaphysical," (Rawls 1999b; 1st pub. 1985).

(under normal circumstances) must not be sacrificed for anything; (b) we judge a society by asking whether it is good for us all, whether it truly is a land of opportunity, and by looking at the quality of life attainable by its nonprivileged members; finally (c) we believe this is what we would choose if we were choosing impartially. Rawls's most central, most luminously undeniable point is that a free society is not a zero-sum game. It is a mutually advantageous cooperative venture. That is why, when given a choice, people almost always choose to live together: They are better off together. Nozick is critical of Rawls, as am I, as are most writers on Rawls. But at the end of the day, isn't there something fundamentally right (even beautiful) in that grand vision?

PUZZLES

1. I just described as "luminously undeniable" Rawls's point that society is not a zero-sum game. Why then do so many people see so much of life as zero-sum? How would you explain the zero-sum mind set?[28]
2. If your university faces a budget crunch, how should you allocate scarce resources? Should you protect existing centers of excellence, cutting budgets of weaker departments? Or, should your budget process work to the greatest advantage of the weakest department (whatever that means in practice)? Why? Is a university's basic structure relevantly unlike a society's?
3. Imagine bargainers in the original position reaching an impasse and agreeing to settle the matter by flipping a coin. Suppose we want to design the thought experiment to yield a desired conclusion, namely that bargainers choose principle X, and so we imagine the coin flip coming up in favor of X. Does this thought experiment justify principle X? Of course not, but why not? The result is not procedurally unfair, so what is the problem?[29] What must we add to the story for it to begin to count as a reason to believe X is a principle of justice?

[28] Here is one answer: People attach importance to income *shares*. Income shares sum to 100%. No share goes up unless someone else's goes down. And there is no such thing as general progress. The sum of all shares never exceeds 100%, so the appearance of stagnation is guaranteed.

[29] I explain why the problem with hypothetical consent arguments gets worse as we try to make them converge on determinate conclusions, in Schmidtz 1990a.

4. Should animals be represented in the original position? If bargainers didn't know what kind of animal they might turn out to be representing, would they decide to arrange inequalities to the greatest advantage of the least advantaged class of animals? If we wanted, we could design the original position so as to convince ourselves that rational bargainers would so decide. Would that give us a reason to regard this principle as a principle of justice? If not, why not?

32

Nozick

THESIS: There is a major problem with what Nozick calls *time-slice* principles; however, not all *patterned* principled are time-slice principles.

HISTORY AND PATTERN

Nozick distinguished historical from patterned principles of justice. The distinction is simple on the surface, but by the time we reach the end of Nozick's discussion, the two categories have become at least three, perhaps four, and not so easily kept separate. Some of Nozick's statements are hard to interpret, but the following is roughly what Nozick intended.[30]

Current Time-Slice principles assess a distribution at a given moment. We look at an array of outcomes. It does not matter to whom those outcomes attach. For example, on an egalitarian time-slice principle, if the outcomes are unequal, that is all we need to know in order to know we have injustice. We do not need to know who got which outcome, or how they got it. History does not matter at all.

End-State principles say something similar, but without stipulating that the outcomes are time slices. So, for example, an egalitarian end-state principle could say we look at lifetime income; if lifetime incomes are unequal, that is all we need to know. The difference between time-slice and end-state principles is this. Suppose again (as per Chapter 22) the Smiths and Joneses have the same jobs at the same factory, but the Joneses are three years older, started working three years earlier, and

[30] I thank Richard Arneson for suggesting how to draw the distinctions. I follow his suggestion to a degree, but not closely enough to make him accountable for the result.

continually get pay raises by virtue of their seniority that the Smiths will not get for another three years. At no time are wages equal, yet lifetime income evens out. We have injustice by an egalitarian time-slice principle, but an end-state principle can look beyond time slices to conclude that the equality required by justice will be achieved.

Patterned principles include both of the above as subsets or examples, but within the broader class are patterns neither time-slice nor end-state. "Equal pay for equal work" is an example of an egalitarian principle that is patterned but neither end-state nor time-slice; it prescribes what outcomes should *track*, in this case the quality and/or quantity of labor inputs, but does not prescribe that outcomes be equal.

Historical principles say what matters is the process by which outcomes arise. Historical principles are complex because, notwithstanding Nozick's intended contrast, patterned principles can have a historical element, and vice versa. "Equal pay for equal work" is both patterned and historical; that is, it prescribes outcomes tracking a pattern of what people have done.

A PROBLEM WITH PATTERNS

Nozick classifies Rawls's difference principle as patterned but not historical (it prescribes a distribution while putting no weight on who produced the goods being distributed.) By contrast, what Nozick calls *entitlement theory* (discussed in the next section) is historical but not patterned.

The problem with patterned principles is that, in Nozick's words, liberty upsets patterns. "No end-state principle or distributional patterned principle of justice can be continuously realized without continuous interference with people's lives."[31] To illustrate, Nozick asks you to imagine that society achieves a pattern of perfect justice by the lights of whatever principle you prefer. Then someone offers Wilt Chamberlain a dollar for the privilege of watching Wilt play basketball.[32] Before we know it, thousands of people are paying Wilt a dollar each, every time Wilt puts on a show. Wilt gets rich. The distribution is no longer equal, and no one complains. Moreover, we are all a bit like Wilt. Every time we earn a dollar, or spend one, we change the pattern. Nozick's question: If justice is a pattern, achievable at a given moment, what happens if you achieve

[31] Nozick 1974, 163.

[32] Nozick 1974, 161–4. Wilt Chamberlain was the dominant basketball player of his era, once (in 1962) scoring a hundred points in a single game.

perfection? Must you then prohibit everything – no further consuming, creating, trading, or even *giving* – so as not to upset the perfect pattern? Notice: Nozick neither argues nor presumes people can do whatever they want with their property. Nozick's point is, if there is *anything at all* people can do – even if the only thing they are free to do is give a coin to an entertainer – then even that tiniest of liberties will, over time, disturb the pattern.[33] It is a mistake, Nozick concludes, to think end-state principles give people what entitlement principles do, only better distributed. Entitlement principles recognize realms of choice that end-state principles cannot recognize. None of the resources governed by end-state principles would ever be at a person's (or even a whole nation's) disposal.[34]

Although Nozick is right in seeing a huge problem with time-slice principles, not all patterned principles are prescriptions for time slices. There are passages where Nozick seems to assume that in arguing against end-state or time-slice principles, he is undermining patterned principles more generally. Not so. Not all patterns are the same, and not all require major interference. Nozick is right that if we focus on time slices, we focus on isolated moments, and take moments too seriously, when what matters is not the pattern of holdings at a moment but the pattern of how people treat each other over time. Even tiny liberties must upset the pattern of a static moment, but there is no reason why liberty must upset an ongoing pattern of fair treatment.

A moral principle forbidding racial discrimination, for example, prescribes no particular end-state. Such a principle is what Nozick calls weakly patterned, sensitive to history as well as to pattern, and prescribing an ideal of how people should be treated without prescribing an end-state distribution.[35] It *affects* the pattern (as would even a purely historical

[33] Nozick 1974, 161–4. See also Feser 2004, 71.

[34] Nozick 1974, 167. Rawls's reply: "The objection that the difference principle enjoins continuous corrections of particular distributions and capricious interference with private transactions is based on a misunderstanding." On the next page, Rawls clarifies: "[E]ven if everyone acts fairly as defined by the rules that it is both reasonable and practicable to impose on individuals, the upshot of many separate transactions will eventually undermine background justice. This is obvious once we view society, as we must, as involving cooperation over generations. Thus, even in a well-ordered society, adjustments in the basic structure are always necessary" (1996, 283–4). The clarification makes it hard to see what Nozick misunderstood. (I thank Tom Palmer for this point.) In any case, one challenge in constructing a constitutional democracy is to limit "necessary adjustments" that signal to citizens that their income is a political football and they are to that extent governed by men, not law.

[35] Nozick 1974, 164.

principle) without *prescribing* a pattern (or more precisely, without prescribing an end-state). And if a principle forbidding racial discrimination works its way into a society via cultural progress rather than legal intervention, it need not involve any interference whatsoever.

If we achieve a society where Martin Luther King's dream comes true, and his children are judged not by the color of their skin but by the content of their character, what we achieve is a fluid, evolving pattern tracking merit rather than skin color. In the process, society comes to require *less* intervention than the relentlessly coercive, segregated society from which it evolved. So, although Nozick sometimes speaks as if his critique applies to all patterns, we should take seriously his concession that "weak" patterns are compatible with liberty. Some may promote liberty, depending on how they are introduced and maintained. The problem is not with patterned principles in general but more specifically with end-state and especially time-slice principles.

A weakness in Nozick's critique of Rawls, then, is this. Nozick is right that time-slice principles license immense, constant, intolerable interference with everyday life, but is Rawls defending such a view? In his first article, Rawls said, "[W]e cannot determine the justness of a situation by examining it at a single moment."[36] Years later, Rawls added, "It is a mistake to focus attention on the varying relative positions of individuals and to require that every change, considered as a single transaction viewed in isolation, be in itself just. It is the arrangement of the basic structure which is to be judged, and judged from a general point of view."[37] Thus, to Rawls, basic structure's job is not to make every transaction work to the working class's advantage, let alone each *member* of the class. Rawls was more realistic than that. Instead, it is the trend of a whole society over time that is supposed to benefit the working class *as a class.* To be sure, Rawls was a kind of egalitarian, but not a time-slice or even end-state egalitarian. The pattern Rawls meant to weave into the fabric of society was a pattern of equal status, applying not so much to a distribution as to an ongoing relationship.[38]

It would be a mistake, though, to infer that Nozick's critique had no point. Nozick showed what an alternative theory might look like, portraying Wilt Chamberlain as a separate person in a more robust sense (unencumbered by nebulous debts to society) than Rawls could

[36] John Rawls, "Outline of a Decision Procedure for Ethics," (1999b, 14; 1st pub. 1951).
[37] Rawls 1971, 87–8.
[38] I thank Alyssa Bernstein for discussion of this point.

countenance. To Nozick, Wilt's advantages are not what Wilt *finds* on the table; Wilt's advantages are Wilt *brings* to the table. And respecting what Wilt brings to the table is the exact essence of respecting him as a separate person.[39]

In part due to Nozick, today's egalitarians are realizing that any equality worthy of aspiration will focus less on justice as a property of a time-slice distribution and more on how people are treated: how they are rewarded for their contributions and *enabled* over time to make contributions worth rewarding. This is progress.

VOLUNTARISM

Nozick says an entitlement theory's principles fall into three categories. First, principles of *initial acquisition* explain how a person or group legitimately could acquire something that had no previous owner.[40] Previously unclaimed land is a historically central example, as are inventions and other intellectual property. Second, principles of *transfer* explain how ownership legitimately is transferred from one person (or group) to another. Finally, principles of *rectification* specify what to do about cases of illegitimate acquisition or transfer.

Nozick favors a version of entitlement theory, grounded in an ideal of voluntarism. Nozick says a distribution is just if it arises by just steps from a just initial position, where the paradigm of a just step is voluntary exchange. As an exemplar of the kind of society that would accord with his brand of entitlement theory, Nozick offers the ideal of a civil libertarian, free market society governed by a minimal state (roughly, a

[39] Of course Rawls wanted the least advantaged to have opportunities do things, not merely experience things. Yet, Rawls's original position embodies a commitment to treat what bargainers have done prior to arriving at the table as arbitrary from a moral point of view. Recall how Chapter 10 likened Rawls's original position to Nozick's experience machine. If we agree that principles of justice must respect what less-advantaged people *do*, we will be agreeing not only to respect what less-advantaged people do *from now on*, on the grounds that respect is in their interest. We will be choosing to respect what they have been doing all along. To give them the respect that is in their interest, we must acknowledge what they deserve, and have deserved all along. Otherwise, our attitude toward them is a patronizing and paternalistic simulation of respect, not the real thing. In effect, Rawls respects our separateness as consumers; Nozick respects our separateness as producers.

[40] Nozick's justification of initial acquisition strikes many as the weak link in his theory, but see Schmidtz 1994.

government that restricts itself to keeping the peace and defending its borders). In such a society, as people interact by consent and on mutually agreeable terms, there will be "strands" of patterns; people amass wealth in proportion to their ability to offer goods at prices that make their customers better off. Employees tend to get promotions when their talents and efforts merit promotion, and so on. However, although society will be meritocratic to that extent, that pattern will be only one among many. There will be inheritance and philanthropy, too, conferring goods on recipients who may have done nothing to deserve such gifts.[41] Is that a problem? Not to Nozick. Nozick joins Rawls in denying that merit is a principle to which distributions (and transfers) must answer. The question, Nozick says, is simply whether people deal with each other in a peaceful, consensual way.

Rawls says, "A distinctive feature of pure procedural justice is that the procedure for determining the just result must actually be carried out; for in these cases there is no independent criterion by reference to which a definite outcome can be known to be just."[42] By this definition, Rawls's is not a theory of pure procedural justice, but Nozick's is. To Nozick, the question is whether proper procedure was followed. So far as justice goes, that is it. There is no other question.

Nozick's theory in a nutshell is that we need not preordain an outcome. We need not know what pattern will emerge from voluntary exchange. What arises from a just distribution by just steps is just. If people want to pay Wilt Chamberlain for the thrill of watching him play basketball, and if this leads to Wilt having more money than the people around him, so be it.

One thing we might say in support of Wilt having a right to live as he does, and our having a right to pay him for doing so, is that Wilt, like us, is a self-owner. After all, it is not for no reason that we speak of people's talents as *their* talents.[43] One way or another, someone makes the call regarding how to use Wilt's athletic skill. Who, if not Wilt, has the right to make that call?

According to G. A. Cohen, liberalism's essence is that people are self-owners. Their lives are their lives; they can live as they please so long as they live in peace. As Cohen defines the terms, right-wing liberals

[41] Nozick 1974, 158.
[42] Rawls 1971, 86.
[43] Feser (2004, 43) deems the argument from self-ownership Nozick's primary argument.

like Nozick believe people can acquire similar rights in external objects, whereas left-wing liberals do not. How should we classify Rawls? Cohen says,

> Rawls and Dworkin are commonly accounted liberals, but here they must be called something else, such as social democrats, for they are not liberals in the traditional sense just defined, since they deny self-ownership in one important way. They say that because it is a matter of brute luck that people have the talents they do, their talents do not, morally speaking, belong to them, but are, properly regarded, resources over which society as a whole may legitimately dispose.[44]

Rawls and Nozick each acknowledged a gap between (1) saying Jane owns her talents and (2) saying Jane owns the cash value of what she produces when she puts her talent to use. Yet, Nozick thought, to take Jane seriously as a separate person is to presuppose her right to make and execute plans of her own, including plans involving the external world. To say Jane can do what she wants, needing our permission *only* if she wants to alter the external world, would be to make a joke of self-ownership. So, where Rawls *embraced* the gap between (1) and (2), Nozick was struggling to *bridge* the gap when he gestured at a theory that we acquire bits of an otherwise unowned external world by working on them.[45] I say "gestured" because Nozick is unsure how the theory works at the edges. Must labor add value? Must labor be strenuous? If I pour my tomato juice into the ocean, why don't I thereby come to own the ocean?[46]

With similar candor, Nozick admits he is not sure what to say about a case where, through no one's fault, our town has only one remaining water source, and Joe is the sole owner of it. The result came about via

[44] G. A. Cohen 2000a, 252. Is this far-fetched? Rawls rejects a "liberal equality" interpretation of the difference principle in favor of a "democratic equality" interpretation (1971, 73ff). Rawls (2001, 75–6) also says "it is persons themselves who own their endowments," which sounds liberal, but then his next remark is, "What is to be regarded as a common asset, then, is the distribution of native endowments, that is, the differences among persons. These differences consist not only in the variation of talents of the same kind (variation in strength and imagination, and so on) but in the variety of talents of different kinds." *Liberal* self-ownership, Cohen might correctly insist, includes the aspects of our selves that distinguish us.

[45] Nozick borrows the theory from John Locke 1960 (*Second Treatise*, chap. 5).

[46] Cohen says, "[T]he claim people can make to the fruits of their own labor is the strongest basis for inequality of distribution, and the claim is difficult to reject so long as self-ownership is not denied" (2000a, 253). In effect, Cohen says, left-liberalism is not viable. Either we deny self-ownership, abandoning left-liberalism in favor of socialism, or embrace inequalities resulting from people freely employing unequal talents, thereby abandoning left-liberalism in favor of right-liberalism. See also G. A. Cohen 2000b. 273–4.

a process that violated no one's rights, we assume, but does that make the result okay? At first, Joe could sell his water for whatever his customers were willing to pay. After Joe accidentally becomes a monopolist, though, Nozick is not so sure. We cannot take any simple concept of property for granted here. We must ask what a community is for, how property rights (and specific ways of establishing legal title, including more or less ritualistic forms of labor-mixing) enable a community to serve its purpose, and how the role of property rights in serving that purpose is superceded in extraordinary conditions. Perhaps, in extraordinary conditions, something else serves that purpose. There are strands of patterns at work here, carrying more weight than Nozick's simple story suggests.

RIGHTS

In Nozick's theory, rights are trumps or *side-constraints*, not merely weights to balance against other considerations. So, what justifies side-constraints? Nozick sometimes is accused of having no foundation, of merely assuming what he needs to prove. For better or worse, though, Nozick borrowed Rawls's foundation, accepting Rawls's premise about the separateness and inviolability of persons, saying, the "root idea, namely, that there are different individuals with separate lives and so no one may be sacrificed for others, underlies the existence of moral side constraints."[47] If this is not a foundation, then Rawls has no foundation either. Nozick's departure was to ask, when we say people are entitled to the most extensive sphere of liberty compatible with a like liberty for all and this right takes priority over other concerns, what if we meant it?

Another answer (to the question of what justifies side-constraints) is that some rights are prerequisites of a society being a cooperative venture for mutual advantage, partly because some rights enable people to know what to expect from each other, and to plan their lives accordingly. On this view, what gives rights their *foundation* also limits their scope. Why do we have rights? Answer: We cannot live well together unless we treat each other as having rights. Why do we have *limited* rights? Answer: We cannot live well together unless we treat our rights as limited. That is how we know Wilt's right to enjoy his property in peace does not include a right to build biological weapons in his garage in an otherwise ordinary neighborhood.

[47] Nozick 1974, 33. See also Lacey 2001, 25ff.

That also is how we know Wilt's right to buy a sports car does not include a right to drive through school zones as fast as he pleases.[48] At some point, a community, guided by a principle that drivers must be allowed to get where they need to go so long as they do not impose undue risk on pedestrians, concludes that anything between ten and twenty miles per hour is within reason, then picks something in that range. After a community posts a limit, such as fifteen miles per hour, drivers no longer have a right to judge for themselves whether twenty is within reason. From that point, pedestrians have a *right* that drivers observe the posted limit.

Such a right is a true side-constraint. Ordinary drivers in normal circumstances have no right to judge for themselves whether the constraint is outweighed. When the constraint applies, it applies decisively. It may, however, have limited scope. For example, the community may decide that the law does not apply to ambulances. If an ordinary driver has an emergency, such as getting his wife to the hospital, and is willing to break laws to get her there sooner, he is liable for the penalty that goes with breaking the law, although courts may at their mercy waive the penalty. If the driver is in an ambulance, by contrast, he does not need the court's mercy. If the law does not apply to ambulances, then he was within his rights.

Suppose you scoff at natural rights. Like Jeremy Bentham, you call them nonsense. Do you still want to live in a society that respects rights? Suppose you investigate, and conclude that societies that respect liberty (that put Rawls's first principle first) are freer, richer, less envious, more open to multicultural enrichment, more respectful of what people produce, and so on. You conclude that you want to live in a society that respects liberty. Is this an argument that we should act as if people have rights, or that people really do have rights? What's the difference?

PUZZLES

1. Nozick uneasily suggests that rights don't stop us from doing what it takes to avoid catastrophic moral horror. Where, then, is the line between overriding rights to avoid catastrophe, and ignoring rights to promote efficiency?
2. Nozick sees forming attachments and working to make the world a better place as parts of what we always have done to make life meaningful. Nozick also thinks we need not compel high mindedness;

48 Nozick 1974, 171.

people left in peace have a history of freely contributing to their communities. Is Nozick right? Under what conditions? What if we are unsure? Nozick may be right: Our neighbors may be as high minded as Nozick hopes. There are no guarantees, though. As a matter of fact, we are unsure. What, if anything, does our uncertainty entitle us to do to each other?

3. As noted, our neighbors cannot build biological weapons in their garages. They cannot drive drunk. Why not? Not because they are hurting us, exactly, but because they are putting us at risk. They are doing the *kind* of thing that *tends* to hurt people.

   Of course, driving within speed limits also imposes risk. The difference is that risks imposed by speeders and drunk drivers are too much, or too pointless. At one extreme, we imagine saying it is okay to shoot at people so long as you do not hit them (no harm, no foul, as they say). At the other extreme, we imagine saying we cannot risk selling hot coffee to people who might spill it on themselves. Few believe either extreme, but where do we draw the line? Should we expect to be able to draw lines with principles of justice? Are principles (as opposed to custom, common law, or evolving community norms) always the right tool for drawing lines?

33

Rectification

THESIS: If it matters how current holdings arose, then justice is to that extent historical, in which case questions arise about how to rectify past injustice. When it is too late to punish wrongdoers or make victims whole, rectification has to be about something other than undoing wrongs.

DOES NOZICK HAVE A THEORY OF JUST DISTRIBUTION? SHOULD HE?

As noted, what Nozick calls entitlement theory incorporates principles of initial acquisition, transfer, and rectification. Nozick's version of entitlement theory embraces an ideal of voluntarism: A distribution is just if it arises by just steps (paradigmatically, voluntary exchange) from a just initial position.

I am not sure Nozick should have said that. It sets the bar high: What can a historical theory say about a world where few titles have an unblemished history? Or perhaps that is simply how it is; like it or not, there is no way from here to a world where distributions are just. Either way, Nozick may have been mistaken in seeing himself as addressing the topic of distributive justice, since his theory wants to go in a different direction. That is, his theory is about justice as how we treat each other rather than justice as cleansing the world's distributions of original sin.

In other words, the core of Nozick's theory is not as previously stated. Nozick's real claim should not be that a distribution is just if it arises by just steps from a just initial position. When Nozick said this, he obscured his real contribution. In truth, Nozick has a theory of just *transfer*, not a theory of just *distribution*: Roughly, a transfer from one person to another is thoroughly just if thoroughly voluntary. The theory ultimately is not so simple, but this is its true essence.

Voluntary transfer cannot cleanse a tainted title of original sin, but any injustice in the result will have been preexisting, not *created* by the transfer. We are fated to live in a world of background injustice, all of us descended from both victims and victimizers, so it is a virtue of Nozick's theory that it does not pretend we might achieve perfect justice if only we can "even the score." Still, it remains possible for moral agents, living ordinary lives, to abide by Nozick's principle of just transfer, and to that (admittedly imperfect) extent to have clean hands.

Nozick says the question, contra patterned principles, is whether a distribution results from peaceful cooperation. More accurately, to avoid encouraging our self-destructive tendency to dwell on histories of injustice, Nozick might have said the question is whether ongoing *changes* in distributions result from peaceful cooperation.

In summary, Nozick meant to offer voluntarism as a basis for a theory of just transfer. He gestured at ideas about how to get unowned resources into the realm of what could be voluntarily transferred. He also gestured at the idea that some part of justice would have to be concerned with undoing wrongful transfers. However, in Nozick's mind, the point of undoing a wrongful transfer is simply to undo a wrongful transfer, not to make current holdings match a favored pattern.

For example, sometimes justice is about returning a stolen wallet to the person from whom it was stolen. Why return the wallet to that person? Not to restore a previously fair pattern but to restore the wallet to the person from whom it was stolen. Sometimes, justice is about *returning* the wallet, not *distributing* it. The wallet's history trumps any thoughts about how it might best be distributed.

FREEDOM AS A ZERO-SUM GAME

G. A. Cohen sees a problem for anyone who, like Nozick, believes in property rights on the grounds that they embody a commitment to peaceful cooperation. This section explains the problem; the next explains what Cohen's insight tells us about the limits of our right to rectify historical injustice.

Cohen's view is that enforcing property rights is as coercive as robbery. Property rights require us not to initiate force. Governments back that requirement with a threat of force, but that very threat is itself an initiation of force.

> I want, let us say, to pitch a tent in your large back garden, perhaps just in order to annoy you, or perhaps for the more substantial reason that I have nowhere to live and no land of my own, but I have got hold of a tent, legitimately or otherwise. If I now do this thing that I want to do, the chances are that the state will intervene on your behalf. If it does, I shall suffer a constraint on my freedom.[49]

To Cohen, "The banal truth is that, if the state prevents me from doing something that I want to do, then it places a restriction on my freedom."[50] His "general point is that incursions against private property which *reduce* owners' freedom by transferring rights over resources to non-owners thereby *increase* the latter's freedom. In advance of further argument, the net effect on freedom of the resource transfer is indeterminate."[51]

There is no denying Cohen's basic point: Even when the state is trying to protect our freedom, its methods are coercive. We would be wrong to infer from this, though, that the net amount of freedom does not change, or even that we will have a problem discerning a change. Cohen's example concerns your garden. What if he were asking not about control of your garden but of your body – about me enslaving you? Would my enslaving you make me more and you less free, with indeterminate net effect?[52] Cohen might agree, the answer is no, but then remind us he was talking about your garden, not your body. Moreover, he never said there is no net effect, only that we would need further argument to discern a net effect. So, if we suppose Cohen's point covers only external goods, such as your title to your garden, what further argument would make the net effect on freedom easier to discern?

Here is a suggestion. What if we treat Cohen's claim not as conceptual analysis but as a testable empirical hypothesis, then compare countries where property titles are stable to countries where they are not?

[49] Cohen 1995, 56.
[50] Cohen 1995, 55.
[51] Cohen 1995, 57.
[52] Cohen sometimes seems to use the word 'free' to refer to (1) an absence of external impediments. This is fine, but there are other kinds of freedom: freedom as (2) an absence of impediments *caused* by other persons; (3) an absence of impediments *deliberately* caused by other persons; (4) an absence of *removable* impediments: that is, impediments not caused by others, but which others have the power to remove; (5) an absence of self-imposed baggage (for example, having made no promises, and being correspondingly free to choose how to spend the rest of one's life). Philosophers argue about which of these is "real" freedom, but the truth is that different senses fit different purposes.

In Zimbabwe, Robert Mugabe and his army are pitching tents wherever they please, and anyone unlucky enough to find Mr. Mugabe in his back garden would rather be elsewhere. No one who knows the unfolding catastrophe that is Zimbabwe could believe that as Zimbabwe's property rights crumble, it merely trades one freedom for another, with indeterminate net effect.

Closer to home, my freedom to drive through a green light comes at a cost of your freedom to drive through a red. Is anything indeterminate about the net effect? Not at all. And property rights manage traffic on our possessions roughly as traffic lights manage traffic on our roads. Both systems help us to form expectations about other people's behavior, and to plan accordingly. A good system of traffic regulation makes everyone more free to go where they wish, even those who currently face red lights. Of course, the red light that some face must turn green from time to time. Further, those who wait must be alert enough to notice when lights turn green. If people are asked to wait forever, or even think they are, the system will break down.

Traffic laws help us to stay out of each other's way. Property laws let us do more; they also help us to engage in trade, with the result that our traffic (our trucking and bartering) leaves fellow travelers not only unimpeded but enriched. The traffic of a healthy economy is a boon, not merely something to tolerate.

Cohen says lack of money is lack of freedom.[53] Cohen also says having money is like having a ticket one can exchange for various things. And, he adds, having such a ticket is a freedom. Let us be clear, though, that on Cohen's analysis, freedom is access to *real* wealth, not merely to pieces of paper offered as a symbol of stored value. A government cannot create more seats in a stadium just by printing more tickets, and likewise cannot create more wealth just by printing more currency. Work creates wealth, and this is not merely a theoretical possibility but is instead our actual history, wherever property rights are stable. If Cohen is right to equate wealth with freedom (and if Cohen is not entirely right about this, he is not entirely wrong either), then a world of stable property rights is not zero-sum. Where property rights are treated with respect, we find that nearly all are wealthier – which is to say, more free in Cohen's sense – than their grandparents were.[54]

53 Cohen 1995, 58.

54 Perhaps I have read Cohen too literally. Perhaps he was not really talking about the indeterminate net effect of the police not preventing you from seizing my back garden

THE LIMITS OF RECTIFICATION

It is in the context of rectificatory justice that Cohen's claim (that protecting property is coercive) has real relevance. When victims and victimizers are long dead, and nothing can be done short of transferring property from one innocent descendent to another, that is when enforcing rights by rectifying an ancient history of unjust transfers really does begin to look like an initiation of force.

Cohen did not intend his point to apply most especially to this aspect of rights enforcement, but this nonetheless is its most poignant and plausible application. Should we enforce property rights of long dead victims at the expense of people who have not themselves initiated force, and who themselves turn out to descend from victims if we go back far enough?[55]

Richard Epstein says, "Any system of property looks backward to determine the 'chain of title' that gives rise to present holdings. But this is not because of any fetish with the past but chiefly from the profound sense that stability in transactions is necessary for sensible forward-looking planning."[56] Dwelling too much on the past is wrong for the same reason that ignoring the past altogether is wrong: Excess in *either* direction reduces stability in transactions, thus making it harder to go forward in peace. A routine title search when buying a house (to verify that the seller's holding of the deed is in fact uncontested) is one thing; going back as many centuries as the land has been occupied is another.

So, if we must return a wallet seized from a previous owner, must we also make sure the previous owner did not seize the wallet in turn? Nozick

for purposes of your own. Perhaps he was talking about arranging an alternative system of property that legalizes or even administers such seizures. Would he have insisted that his favored system, backed by the threat of force, would be as coercive as robbery? I do not know.

[55] Chandran Kukathas says people cannot be held responsible for injustices committed before they were born; they are not at fault. Neither are societies responsible. Societies do not choose or act, so are not the kind of entity that can be at fault. Yet, rejecting all accountability for the past comes close to rejecting justice itself. So, Kukathas argues, we need a third option, and we have one. Even if people now living are blameless, associations to which some of them belong may be held responsible because *they did the deeds*. "Without going into the details of Aboriginal history since settlement, it is enough to note that there is more than sufficient evidence to confirm that many injustices were committed by the governments of the various states of Australia. To a lesser extent, the sins of the church are also on the record. In these circumstances, the attribution of responsibility for past injustice is not a problem: It can be laid plainly at the door of those associations, still in existence, which committed them" (2003, 183).

[56] Epstein 2003, 130.

envisions a civil libertarian society where we carry on from where we are, in peace. Yet we cannot carry on in peace unless there is a limit on our obligation to undo the past. So, what sort of limit would be philosophically respectable?

THOSE WHO CANNOT FORGET THE PAST ARE DOOMED TO REPEAT IT[57]

There are places where people have been "evening the score" for centuries, and the cycle of mutual destruction will not stop until people *aim* to leave the past behind. Linda Radzik says true rectification is about victims and victimizers (or their descendants) repairing relationships and setting the stage for a peaceful future.[58]

The point is to make amends, but for making amends to succeed, there has to be uptake: Victims and their descendents, for their own sakes, have to embrace the aim of achieving closure. Descendants of victims, for their own sakes, must accept that guilt is not a weapon to be used against a perpetrator's descendents forever.

There is no future in evening the score. We have histories of uncorrectable injustice, scores never to be settled. When vengeance is taken against innocent descendants, every act of vengeance becomes another score needing to be settled against someone else's innocent descendants.[59]

In another sense, though, settling a score is indeed feasible: Namely in Radzik's sense, where we conceive of settling a score not as undoing wrongful transfers nor as taking revenge but as making amends. A cycle of revenge inflicted on innocent descendants cannot end until people aim at something different. Once people aim at making amends, though, they *are* aiming at something different, something with a chance of setting the stage for a more peaceful future.

It is as Mr. Mandela says. "We can have prosperity, or we can have revenge. But we cannot have both. We need to choose."[60] This suggests where we might find a limit on our duty to undo the past. When we ask how far back we should go, the schematic yet instructive answer is: As far

57 The title is borrowed from a quip by Brian Barry 2005, 254.
58 Radzik 2004.
59 Sher 1997 argues that claims of ancient injustice must fade with time.
60 Several people quoted this statement in conversation during my first visit to South Africa in 1999.

back as we need to go to repair broken relationships. The aim is to achieve closure among people who aim to achieve closure. (Not everyone does, of course, so we also need to make sure we play no part in letting people become addicted to playing the guilt card. Successful rectification is not a crutch.) As Charles Griswold puts it, our aim is to live constructively in a world that we acknowledge is profoundly marred; part of what it means to *reconcile* is to come to terms with life in an imperfect world.[61]

On a view like Radzik's, the point of undoing wrongful transfers is not to even the score but to bring closure to a history of wrongful transfers. Thus, South Africa's Truth and Reconciliation Commission set out in 1995 to document human rights abuses between 1960 and 1994. Part of its mandate is to grant amnesty to those who cooperate in documenting relevant facts. Now, these crimes were not ancient. It was not a situation where innocent people were being asked to pay for crimes of their ancestors. Many of apartheid's perpetrators were very much alive, and by no means beyond the reach of the law. Yet, even so, Mandela's goal (like Desmond Tutu's) was reconciliation, not revenge. He wanted to prevent the legacy of apartheid from continuing to hang over future generations.

We set the stage for carrying on by finding what happened, acknowledging, mourning, then vowing to do all we can to make sure history never repeats itself. The future matters even when the past cannot be undone. There is no injustice in being willing to carry on.[62] Or even when something is not quite right with carrying on, it is a lesser evil than an endless cycle of vengeance.

Another example: Innocent Japanese Americans were imprisoned during World War II. Someone had to try to make amends. In turn, Japanese American victims (and their descendents) had to decide what sort of gesture to accept. In deciding what to accept, they had to be sensitive to the degree to which people offering the gesture were not guilty, but merely represented the guilty, perhaps only by virtue of being of the same race. When President Reagan signed the Civil Liberties Act of 1988, appropriating $1.25 billion (augmented by President Bush) for reparations to internees and their descendents, then in 1999 when President Clinton formally apologized on behalf of the nation and broke ground on a national memorial, it was too late and the crime too huge for anything to

[61] Personal communication, 2003.

[62] I thank Chris Griffin and Cindy Holder for conversations on this topic.

make victims whole,[63] but victims or their descendents still had to decide whether to accept the gesture. All parties had to shift focus away from compensating for the past and toward healing an ongoing relationship, establishing a base for mutual respect going forward, then getting on with their lives.

[63] As Jeremy Waldron observes, the point of the money "was to mark – with something that counts in the United States – a clear public recognition that this injustice did happen, that it was the American people and their government that inflicted it, and that these people were among its victims" (1992, 7).

34

Two Kinds of Arbitrary

THESIS: Historical injustice calls for rectification, but arbitrariness in natural distributions is not unjust, and does not need rectifying.

WHEN DO WE HAVE A RIGHT TO DISTRIBUTE?

Nozick thought a bias against respecting separate persons lurks in the very idea of *distributive* justice. The idea leads people to see goods as having been distributed by a mechanism for which we are responsible. Nozick believes there generally is no such mechanism and no such responsibility. "There is no more a distributing or distribution of shares than there is a distributing of mates in a society in which persons choose whom they shall marry."[64]

The lesson: If we have a license to distribute X, then we ought to distribute X fairly, and Rawls gave us a theory about how to do that. However, we lack a license to distribute mates. Thus, we have no right to distribute mates unfairly, and neither do we have a right to distribute mates fairly. They are not ours to distribute.

[64] Nozick 1974, 150. I thank Jerry Gaus for reminding me of the following remark by David Gauthier: "If there were a distributor of natural assets, or if the distribution of factor endowments resulted from a social choice, then we might reasonably suppose that in so far as possible shares should be equal, and that a larger than equal share could be justified only as a necessary means to everyone's benefit. . . . In agreeing with Rawls that society is a cooperative venture for mutual advantage, we must disagree with his view that natural talents are to be considered a common asset. The two views offer antithetical conceptions of both the individual human being and society" (1986, 220–1).

What about inequalities? The same point applies. Unless a particular inequality is ours to arrange, theories about what would be fair are moot. More generally, to show that I have a right to distribute X according to a given plan, we may at some stage need to show that my plan is *fair*, but before that, we need to show that X's distribution falls within my jurisdiction. Thus, in effect, Rawls's principles do not start at the start. Rawls's principles tell us how to distribute X, given that the distribution of X is our business. But the latter is not a given.

THE WORD "ARBITRARY" HAS TWO MEANINGS

Rawls speaks of mitigating the arbitrary effects of luck in the natural lottery.[65] Is there a difference between a lottery Jane wins by luck of the draw, and a lottery rigged to make sure Jane wins? Is there a difference between Joe turning out to be less skilled than Jane, versus a situation where Joe deliberately is held back to *make sure* Jane will be more skilled? As one way of motivating his two principles, Rawls says, "Once we decide to look for a conception of justice that *nullifies* the accidents of natural endowment and the contingencies of social circumstance . . . we are led to these principles. They express the result of leaving aside those aspects of the social world that seem arbitrary."[66]

Arbitrary? The word has two meanings. Natural distributions can be arbitrary, meaning *random.* Or choices can be arbitrary, meaning *capricious.* In one case, no choice is made. In the second, an unprincipled choice is made.[67] There is a difference. In fair lotteries, winners are chosen at random. A *rigged* lottery is unfair. Why? Because it *fails* to be arbitrary in the benign sense. It is by *failing* to be arbitrary in the benign sense that it *counts* as arbitrary in the bad sense. What of the natural lottery, then? The natural lottery is arbitrary in the benign sense, but how does that connect to being unfair in the way capricious choice is unfair?

It doesn't. Rawls says, "Intuitively, the most obvious injustice of the system of natural liberty is that it permits distributive shares to be improperly influenced by these factors so arbitrary from a moral point of view."[68]

[65] Rawls 1971, 74–5.

[66] Rawls 1971, 15, emphasis added.

[67] When we call a choice arbitrary, we imply not only that it is unjustified, not only that it is wrong, but that it exhibits a certain arrogance: For example, a person's attitude might be "I can do what I want."

[68] Rawls 1971, p. 72.

However, when 'arbitrary' means random, as it does in this passage, there is no connection between being arbitrary and being improper. Capricious choice wears impropriety on its sleeve; the natural lottery does not. Had Joe's mother assigned Jane all the talent, deliberately leaving Joe with none, we might at least wonder why. In fact, though, Joe's mother did not assign him less talent. It just happened. It was chance, not caprice.

Put it this way: Life is about playing the hand you are dealt. Being dealt a bad hand is not the same as facing a stacked deck. A deck is stacked only if a dealer deliberately stacks it, declining to leave the matter to chance.

THE NATURAL LOTTERY

Rawls says, "We are led to the difference principle if we wish to arrange the basic social structure so that no one gains (or loses) from his luck in the natural lottery of talent and ability, or from his initial place in society without giving (or receiving) compensating advantages in return."[69] Interestingly, Rawls took the trouble to put "or loses" in parentheses, signaling that in his mind gaining is a problem all by itself. But if a life-extending mutation appears in a population, should we arrange basic structure so no one gains from the mutation? No. Gaining is good. There is a problem if Jane gains at Joe's expense, but then the problem is still with the losing, not the gaining.

To Rawls, "[I]t is not just that some should have less *in order* that others may prosper."[70] I agree. Yet, as noted, a natural lottery is not a stacked deck; moreover, if someone *had* stacked the deck, deliberately assigning Joe a lazy character and no talent, the reason would not be so "others may prosper." Joe being untalented is no help to others. On the contrary, to assign Joe a talent level that would help others prosper, we would need to assign Joe *more* talent, not less. Making Joe a provider of high-quality services would help others prosper, thus providing a real reason to compensate him.

One way (the only way I know of) to rationalize the idea that *Jane's* being more talented entitles *Joe* to compensation is to suppose that life is like a zero-sum poker game in which the more talented Jane is, the less chance Joe has of winning. If Jane is more talented, Jane captures more

[69] Rawls 1999a, 140. Strictly, this consideration leads only to the Precursor. We get the difference principle only after deciding to "complete" the theory, then rejecting other ways of completing it.

[70] Rawls 1971, 15, emphasis added.

pie, and captures more at Joe's expense. However, it is Rawls's point, after all, that society is not a zero-sum card game, but a cooperative venture in which the pie's size is variable. Almost all people can have a better life than they could have had on their own, and the reason is simple: Other people's talents make all of us better off. Talented bakers don't *capture* pie. They *make* it.[71] The rest of us have more pie, not less, when talented people put their talent to work.

The natural lottery is not zero-sum. When a baby is born with a cleft palate, it is not "in order that others may prosper." When the next baby is born healthy, needing no special care, this baby's health does not come at the first baby's expense.[72] Rawls says it is unjust that some should have less so that others may prosper, but the first baby does not have a cleft palate so that the second baby may prosper.

Rawls says, "The natural distribution is neither just nor unjust; nor is it unjust that persons are born into society at some particular position. These are simply natural facts. What is just and unjust is the way that institutions deal with these facts."[73] If Rawls is right, then when institutions "deal with natural facts," they are not undoing wrongs.

THE REAL ISSUE

A distribution of talent per se is not a problem, solvable or otherwise. But even if, as Rawls says, there is no injustice in a natural distribution, there may yet be a problem. Being born with a cleft palate is a problem. The problem is not that a cleft palate is unjust but that it is bad. Its badness gives us some reason to intervene to fix the problem.

But note the real issue: We are not trying to fix an *improper distribution* of cleft palates. We are trying to fix cleft palates.

[71] Needless to say, even the most self-reliant bakers get help. See Chapter 16.

[72] If others do complain, then as Cohen says about appropriating water in a world of plenty, "[Y]our powerful reply is to say that no one has any reason to complain about your appropriation of the water, since no one has been adversely affected by it" (1995, 75). See also Wellman 2002, 66.

[73] Rawls 1971, 102.

35

Procedural versus Distributive Justice

THESIS: We can reconsider Rawls's original position, asking how we might have designed it if we really thought we had no independent criterion for the right result.

GETTING TO THE DIFFERENCE PRINCIPLE

If we imagine bargainers behind a veil of ignorance, not knowing what position they will occupy, that makes the situation fair, but (as Rawls knew) what suffices to make a situation fair is not enough to entail that bargainers choose the difference principle. According to pure procedural justice as Rawls defines it, this should be immaterial, since "pure procedural justice obtains when there is no independent criterion for the right result: instead there is a correct or fair procedure such that the outcome is likewise correct or fair, whatever it is, provided that the procedure has been properly followed."[74]

Yet, as Rawls acknowledged, he did desire to reach a particular conclusion, so he made further assumptions to make an otherwise pure procedure converge on what he considered the right result. For example, bargainers not only do not know what position they occupy; they do not know what skills they possess, or what skills their society prizes. They do not know what they personally believe about justice, and so have no basis for choice other than their calculation of what is in their interest. They do not know what probabilities to attach to prospects of being at the minimum, and so have no basis for discounting improbable risks. Bargainers

[74] Rawls 1971, 86.

do know they are choosing for a closed society. No one enters except by birth, or leaves except by dying.[75]

Rawls speaks of bargainers being "directed" by a maximin rule (according to which one chooses the path whose worst possible outcome is better than the worst outcomes of the alternatives). Why are they "directed" by maximin? Rawls's answer: Because they do not care what they gain above the minimum, so long as they know:

1. the minimum provided under a maximin rule would be "completely satisfactory," and
2. the minima of societies under alternatives to maximin are significantly below that level and may be altogether intolerable.[76]

The more assumptions we add, though, the worse the strains of commitment when we lift the veil and return to this world.[77] For example, if society would have to be closed for the difference principle to be a rational choice, and if in fact we live in open societies from which talented people exit when they can do better elsewhere, where does that leave the difference principle?[78] As Rawls says, bargainers "should not reason from false premises. The veil of ignorance does not violate this idea, since an absence of information is not misinformation."[79] Two points: First, the veil not only *deprives* people of knowledge they have in the real world; it *endows* them with knowledge (for example, of conditions such as 1. and 2., above) that no real bargainer has. Second,

[75] Rawls 1996, 12. Rawls says the closed-society assumption lets us focus on certain questions free from distracting details, adding that he will discuss justice between "peoples" later. Justice between "peoples" is not the truly distracting detail, though. The *distracting* detail is that in open societies, talented people *emigrate* when staying becomes disadvantageous.

[76] Rawls 2001, 98.

[77] Rawls cautions us that the original position is "not a gathering of all actual or possible persons. To conceive of the original position in either of these ways is to stretch fantasy too far; the conception would cease to be a natural guide to intuition" (1971, 139). Why? "Stretching" the fantasy in this way would not make it any less realistic than it already is, so what is the problem? How would it cease to be a natural guide to intuition? Which intuition? The real problem: Among "all actual or possible persons" are quadriplegics, and Rawls's intuition is that justice is about arranging inequalities to the greatest advantage not of quadriplegics but of the working poor.

[78] John E. Roemer says, "The benefit of the veil of ignorance construct is that it forces objectivity, or impartiality. But the cost is that we must make decisions with a great handicap – we have discarded massively important information that is available to us in the real world" (2002, 183).

[79] Rawls 1971, 153.

some of that information (for example, that society is closed) is indeed misinformation.[80]

WHY CLOSED SOCIETY?

It is curious that Rawls theorized about justice in closed societies, since as a matter of historical record the least advantaged have always been better off in open societies, societies where people are free to move in search of better opportunities. If we are theorizing about what kind of society is best for the least advantaged – if that is the desired conclusion – then is anything more fundamental than freedom of movement? Indeed, why not deem freedom of movement the core of the *first* principle: Everyone has a right to live in a maximally open society, a society where they have no obligation to stay if they would rather be elsewhere?

GETTING THE DESIRED SOLUTION

As with other issues of Rawls interpretation, there is no consensus regarding what Rawls had in mind when he repeatedly said, "We want to define the original position so that we get the desired solution."[81] Did he mean we define the original position to get his two principles?[82] In any case, the point is not that the project is illegitimate. Rawls's original position may still be a fair (if not uniquely fair) test of competing conceptions, and Rawls's two principles may be capable (if not uniquely capable) of passing that test.

How would we decide whether the original position is a fair test? (A different question: How would we decide whether the original position is testing for fairness?) Consider that the original position would, if real, put bargainers in a position of not being able to bias negotiations in their own favor. Not knowing what position they occupy forces real bargainers to negotiate impartially. But if *this* feature is what marks the original position as fair – as I believe – then other features (for example the assumption that society is closed) are altogether dispensable. At least they would be altogether dispensable if our only goal were to "set up a fair procedure so that any principles agreed to will be just."[83]

[80] Likewise in the category of massively important misinformation is the idea that bargainers generally do not care what they gain above the minimum.

[81] Rawls 1971, 141.

[82] Special thanks to David Estlund and Alex Kaufman for helping me see different aspects of this issue.

[83] Rawls 1971, 136.

SUPPOSE WE DON'T KNOW WHAT CONCLUSION WE DESIRE

So, suppose there is no "desired solution" and we seek simply to articulate an ideal of pure procedural justice, without trying to make the procedure track any independent standard. Suppose we simply want to preserve the impartiality (thus the fairness) of the original position. What procedure might we set up?

Here is a suggestion. Suppose we theorists put *ourselves* behind a veil of ignorance. Imagine us trying to construct a fair bargaining game *without knowing our own conception of justice.* Suppose we do not *know* what solution we desire, and thus, unlike Rawls, cannot "define the original position so as to get the desired solution." What would we do? Would we posit that bargainers do not care who brought what to the table; they care only about what they get, yet care little for what they might get above the minimum?[84] Would we posit grave risks in seeking gains above the minimum? Would we assume disabled people are not represented at the bargaining table? Would we imagine ourselves picking rules for a closed society? Would we assume bargainers start with equal claims to (the distribution of) each other's advantages, conceived as a common asset?

I think we would do none of the above. We might assume bargainers do not know their position in the distribution, since intuitively that has something to do with impartiality, which intuitively has to do with fairness. But if we do not know our conception of justice – if for all we know we are as likely to be elitists as to be egalitarians – then we will not design the situation to converge on a particular conception. We would expect bargainers in that setting to try to choose principles that were good for all, but would have no reason to predict anything more specific. We would *expect* different groups of bargainers to converge on different conclusions.

Moreover, that would not trouble those who take seriously the idea of procedural justice. On a procedural conception, we conclude that if, after fair deliberation, people agree to bind themselves to each other in a particular way, then by that very fact they *are* bound to each other in that way. If other people, after similarly fair deliberation, agree to bind themselves on different terms, then they *are* bound by those different terms. By the lights of procedural justice, the idea that different groups of people could be bound in different ways (and that philosophical theorizing is

[84] As Thomas Nagel (1989, 12) worries, "Keeping in mind that the parties in the original position do not know the stage of development of their society, and therefore do not know what minimum will be guaranteed by a maximin strategy, it is difficult to understand how an individual can know that he 'cares very little, if anything, for what he might gain above the minimum.'"

not enough to tell us how particular people are bound – we need to know their actual histories) is not a problem.

NOT KNOWING WHETHER EQUAL SHARES IS THE DEFAULT

Suppose some bargainers say, "We didn't come to the table to talk about how to distribute the stuff on the table. We came because the stuff on the table is ours. We came here to reclaim it." Would a thought experiment like mine be relevant to a world where people have prior claims to the goods on the table? Maybe not, and maybe that is a good objection. But in that case, we must conclude not that we should reject *this* thought experiment but that we should reject *all* such thought experiments. All such experiments assume we can focus on distributing goods as if goods were presenting themselves to us more or less in an unowned state – as if we were at liberty to distribute them in any manner we deem fair. If that assumption is wrong, then all such thought experiments are wrong.

I wish I knew of a variation on the theme of the original position with three advantages:

1. My ideal original position would avoid giving equal shares or any other distribution a position of unearned privilege in a debate about how the distributing ought to go.
2. My ideal original position would avoid preconceptions about the *range* of goods bargainers are entitled to distribute. It would not assume bargainers are arranging a distribution of talents or inequalities or mates but would instead assume bargainers gather to distribute whatever is as yet unclaimed. Ideal bargainers would not assume they have a right to distribute goods that have histories of belonging to someone else. They might learn in particular cases that an item's history is like that of a wallet that was stolen and should be returned to a prior owner, but they would see that what they are doing in such cases is undoing wrongful transfers, not distributing.
3. Along the same lines, my ideal original position would untangle issues of distributive and rectificatory justice. Rawls says the principle of redress holds, "that undeserved inequalities call out for redress. . . . The idea is to redress the bias of contingencies in the direction of equality."[85] Rawls adds, "[W]hatever other principles

[85] Rawls 1971, 100.

we hold, the claims of redress are to be taken into account."[86] Not so. No one accepts what Rawls calls the principle of redress unless they already accept that undeserved inequalities are unjust (and that redress consists of moving from undeserved inequality in the direction of undeserved *equality* rather than, say, *deserved* inequality). We need to settle that justice requires X before we have any license to say departures from X call for redress.[87] In the original position, though, bargainers are supposed to be *deciding* what calls for redress. Undeserved inequality? Undeserved *equality*? Or nonconsensual transfer?

I cannot think of a version of the original position with all these advantages, but any version lacking them is question begging in one way or another.[88] Someone may one day devise a version of the original position with these virtues, but until that day, I am predicting that further progress in political theory will have nothing to do with original position thought experiments.

AN INCOMPLETE THEORY IS WHAT WE NEED

Recall the precursor: An inequality is allowed only if the institution allowing it works to everyone's advantage. As noted, Rawls worried that such a principle is incomplete. We considered how Rawls proceeded to complete it.

But why do we need a theory to be "complete?" We might suspect that all we need is a theory saying what is unjust, and that the precursor is

[86] Rawls 1971, 101. Some say the principle of redress *is* the difference principle. Rawls says not, but says the fact of its giving the redress principle some weight is a point in the difference principle's favor.

[87] Rawls says, "The natural distribution is neither just nor unjust; nor is it unjust that persons are born into society at some particular position. These are simply natural facts. What is just and unjust is the way that institutions deal with these facts" (1971, 102). So, if a distribution is not unjust, what needs correcting? Earlier in the same passage, Rawls says, "the difference principle gives some weight to considerations singled out by the principle of redress. This is the principle that undeserved inequalities call for redress" (Ibid). Is it crucial that the inequalities calling for redress are undeserved? If not, then why not simply say *inequalities per se* (along certain dimensions?) call for redress? Or, if the notion of desert is pivotal, then do undeserved *equalities* likewise call for redress?

[88] Rawls was not question begging when he showed that if we *do* grant equality a position of unearned privilege, we *still* end up rationally departing from equality. We beg the question when we use the "equal shares is a default" premise in arguments *for* equality, which is how most people use it today.

complete enough for that. It rules out sacrificing people for the general good, thus enshrining respect for the separateness of persons.

Indeed, we might conclude we do not *want* a "complete" theory. When we ask what we want from basic structure, we realize we *need* incompleteness.

Or perhaps the point is that we need principles, not only rules, but we mistakenly think a principle is incomplete until converted from a principle into a rule. In any case, Rawls sketches four ways of "completing" a theory and says bargainers would choose his difference principle, but never acknowledges that choosing among the four may not be a theory's job, and indeed may not be a basic structure's job. Perhaps any of the four, freely chosen by people to be governed by it, would qualify as just. (Rawls acknowledges this in the international arena; we neither need nor want a complete theory, for we view self-determination as a fundamental good for "peoples.")[89]

A basic structure's job is to get a political community off the ground, enabling voters and legislators to define and refine community norms as they go. Communities whose basic structure evolves toward completion in ways that benefit everyone, as the precursor requires, will be just, more or less. The final sentence of Jon Mandle's overview of Rawls says, when "the design of the basic structure is at stake," citizens rely "on principles that all reasonable people can share."[90] Because the precursor is more open ended than the difference principle, it is closer to being what reasonable people can be expected to share, letting contractors launch a community while deferring ongoing elaboration to legislators and voters. In any case, getting from the precursor to a full-blown difference principle is not easy. We can imagine something along the lines of the precursor being part of an overlapping consensus among real-world reasonable people. We cannot say the same of the difference principle.[91]

[89] Rawls 1999c, 85.

[90] Mandle 2000, 151.

[91] David Miller reports on laboratory experiments where groups of five subjects were put in situations designed to mimic the veil of ignorance. Given a choice, 4% of individuals, and *no* groups, chose maximin from a list of options. Nearly one quarter chose to maximize average payoff; three quarters chose to maximize the average subject to a guaranteed minimum. Groups wanted gains above a minimum to be distributed in proportion to subjects' contributions. Concern for the least well off "expressed itself in support for an income floor rather than for the difference principle" (1999, 80–1). Miller says subjects preferred that even the floor not be unconditional. Subjects wanted people to do what they could, not only to earn rewards above the floor, but to qualify for the floor itself.

Rawls sometimes says all we do at a level of theory is pick a framework; societies work out details.[92] This is the right thing to say. Most of what makes a society liberal cannot be guaranteed by basic structure but is instead in the hands of people and communities working out their own destinies within basic structures.[93]

FINAL THOUGHT

Theorizing about justice aims to articulate principles of justice, but principles of justice are principles, not rules. Rules are meant to *determine* our thinking about what to do; principles are meant to *guide* it. At the normative level, my theory says there are four basic types of principle, perhaps more, that there is no one of them to which the others can be reduced, and that they do not add up to a decision procedure.

Theories are not merely somewhat like maps. They are a lot like maps. One implication: There is no exact truth of the matter about how a theory of justice ought to look: whether a theory at the normative level ought to specify two principles, or four elements, or to what extent the principles resemble a decision procedure.[94] There is a truth about the terrain being mapped, but when the question is how to *represent* the terrain, the answer is that we have to choose. It is possible to make mistakes – to represent the terrain in misleading or unhelpful ways. However, there is no uniquely right way to do it. Moreover, *any* representation is at least potentially misleading or unhelpful.

A map is a leap of faith and an act of trust on the part of a mapmaker, as is any attempt to communicate with readers. A map can be true in the sense of being apt for giving users (assumed to have good faith and some map reading skills) an impression of the terrain accurate enough for the purposes they bring to the map. Such truth is not in the map per se, though, but is instead an interface between map and user.

[92] Rawls 1996, 377.

[93] See Tomasi 2001 for the most developed version of this idea.

[94] I owe some of this thought to Jason Brennan. Brennan 2005 classifies my theory as, in his terms, global antirealist but local realist, meaning there are ground facts about justice, but no overarching and unique truth about how a theory should represent those ground facts.

References

Ackerman, Bruce A. 1980. *Social Justice in the Liberal State.* New Haven, CT: Yale University Press.

Ackerman, Bruce A. 1983. "On Getting What We Don't Deserve," *Social Philosophy & Policy* 1: 60–70.

Anderson, Elizabeth S. 1999. "What Is the Point of Equality?" *Ethics* 109: 287–337.

Aristotle. *Politics,* Book III, Chap. 12, 1282b.

Arneson, Richard J. 1999. "Arneson on Anderson," Brown Electronic Article Review Service (BEARS). http://www.brown.edu/Departments/Philosophy/bears/9904arne.html

Arneson, Richard J. 2003. "Equality, Coercion, Culture, and Social Norms," *Politics, Philosophy, and Economics* 2: 139–63.

Arrow, Kenneth J. 1971. "A Utilitarian Approach to the Concept of Equality in Public Expenditures," *Quarterly Journal of Economics* 85: 409–15.

Baker, Edwin. 1974. "Utility and Rights: Two Justifications for State Action Increasing Equality," *Yale Law Journal* 84: 39–59.

Barry, Brian. 2005. *Why Social Justice Matters.* Cambridge: Polity.

Becker, Lawrence C. 1980a. "The Obligation to Work," *Ethics,* 91: 35–49.

Becker, Lawrence C. 1980b. "Reciprocity and Social Obligation," *Pacific Philosophical Quarterly* 61: 411–21.

Becker, Lawrence C. 1986. *Reciprocity.* New York: Routledge & Kegan Paul.

Becker, Lawrence C. 1998. "Afterword: Disability, Strategic Action, and Reciprocity," in Silvers, Wasserman, and Mahowald, eds. *Disability, Difference, Discrimination.* 293–303. Lanham, MD: Rowman and Littlefield.

Becker, Lawrence C. 2003. "Reciprocity (But I Repeat Myself)," unpublished manuscript. Oral Presentation to the Virginia Philosophical Association.

Beitz, Charles. 1979. *Political Theory and International Relations.* Princeton, NJ: Princeton University Press.

Boskin, Michael, Ellen Dulberger, Robert Gordon, Zvi Griliches, and Dale Jorgensen. 1996. "Toward a More Accurate Measure of the Cost of Living:

Final Report to Senate Finance Committee," currently available at www.socialsecurity.gov.
Braybrooke, David. 1987. *Meeting Needs.* Princeton, NJ: Princeton University Press.
Brennan, J. 2004. "Rawls's Paradox," University of Arizona, unpublished manuscript.
Brennan, J. 2005. "The Best Moral Theory Ever," University of Arizona, unpublished manuscript.
Brock, Gillian. 1999. "Just Deserts and Needs," *Southern Journal of Philosophy* 37: 165–88.
Broome, John. 1991. *Weighing Goods: Equality, Uncertainty, and Time.* Oxford: Blackwell.
Buchanan, Allen. 1990. "Justice as Reciprocity versus Subject-Centered Justice," *Philosophy and Public Affairs* 19: 227–52.
Burtless, Gary. 1990. *A Future of Lousy Jobs? The Changing Structure of U. S. Wages.* Washington, DC: Brookings Institute.
Carter, Alan. 2001. "Simplifying Inequality," *Philosophy and Public Affairs* 30: 88–100.
Cohen, Andrew Jason. 1999. "Communitarianism, Social Constitution, and Autonomy," *Pacific Philosophical Quarterly* 80: 121–35.
Cohen, Andrew Jason. 2004. "What Toleration Does and Does not Require From Liberalism," unpublished manuscript.
Cohen, G. A. 1995. *Self-Ownership, Freedom, and Equality.* Cambridge: Cambridge University Press.
Cohen, G. A. 2000a. "Self-Ownership, World-Ownership, and Equality," in Vallentyne and Steiner, eds. *Left-Libertarianism and Its Critics.* 247–70. New York: Palgrave.
Cohen, G. A. 2000b. "Self-Ownership, World-Ownership, and Equality," in Vallentyne and Steiner, eds. *Left-Libertarianism and Its Critics.* 271–89. New York: Palgrave.
Cowen, Tyler. 1998. *In Praise of Commercial Culture.* Cambridge, MA: Harvard University Press.
Cowen, Tyler. 2000. *What Price Fame?* Cambridge, MA: Harvard University Press.
Cox, W. Michael and Richard Alm. 1995. "By Our Own Bootstraps," Federal Reserve Bank of Dallas annual report.
Christiano, Thomas. 2005. *The Constitution of Equality.* New York: Oxford University Press.
Daniels, Norman. 1978. "Merit and Meritocracy," *Philosophy and Public Affairs* 7: 206–23.
Duncan, Greg, Johanne Boisjoly, and Timothy Smeeding. 1996. "How Long Does It Take for a Young Worker to Support a Family?" Northwestern University Policy Research Website.
Easton, Loyd D. and Kurt H. Guddat (eds.) 1967. *Writings of the Young Marx on Philosophy and Society,* Garden City, NY: Anchor Books.
Epstein, Richard. 2003. *Skepticism and Freedom.* Chicago: University of Chicago Press.
Feinberg, Joel. 1970. *Doing & Deserving.* Princeton, NJ: Princeton University Press.
Feinberg, Joel. 1984. *Harm to Others.* New York: Oxford University Press.

Feldman, Fred. 1995. "Desert: Reconsideration of Some Received Wisdom," *Mind* 104: 63–77.
Feser, Edward. 2004. *On Nozick*. Toronto: Wadsworth.
Folbre, Nancy and Julie A. Nelson. 2000. "For Love Or Money – Or Both?" *Journal of Economic Perspectives* 14: 123–40.
Foot, Philippa. 1967. "The Problem of Abortion and the Doctrine of Double Effect," *Oxford Review*. 5: 5–15.
Frankfurt, Harry. 1987. "Equality as a Moral Ideal," *Ethics* 98: 21–43.
Fried, Barbara H. 2005. "Begging the Question With Style: *Anarchy, State, and Utopia* at Thirty Years," *Social Philosophy and Policy* 22: 221–54.
Galbraith, James K. 2000. "Raised on Robbery," *Yale Law & Policy Review* 18: 387–404.
Galston, William A. 1980. *Justice and the Human Good*. Chicago: University of Chicago Press.
Gaus, Gerald F. 2000. *Political Concepts and Political Theories*. Boulder, CO: Westview Press.
Gauthier, David. 1986. *Morals By Agreement*. Oxford: Oxford University Press.
Gottschalk, Peter and Sheldon Danziger. 1999. "Income Mobility and Exits from Poverty of American Children, 1970–1992," Boston College Working Papers in Economics Website, 430.
Griffin, James. 1986. *Well-Being: Its Meaning, Measurement, and Moral Importance*. Oxford: Clarendon Press.
Hare, R. M. 1982. "Ethical Theory and Utilitarianism," in Sen and Williams, eds. *Utilitarianism and Beyond*. 23–38. Cambridge: Cambridge University Press.
Harman, Gilbert. 1998. "Ethics and Observation," in Geoffrey Sayre-McCord, ed. *Essays on Moral Realism*. 119–24. Ithaca, NY: Cornell University Press.
Harsanyi, John C. 1955. "Cardinal Welfare, Individualistic Ethics and Interpersonal Comparisons of Utility," *Journal of Political Economy* 63: 309–21.
Hart, H. L. A. 1961. *The Concept of Law*. Oxford: Clarendon Press.
Hayek, F. A. 1960. *The Constitution of Liberty*. Chicago: University of Chicago Press.
Hinderaker, John H. and Scott W. Johnson. 1996. "Wage Wars," *National Review* April 22: 34–8.
Holmgren, Margaret. 1986. "Justifying Desert Claims: Desert and Opportunity," *Journal of Value Inquiry* 20: 265–78.
Hubbard, R. Glenn, James R. Nunns, and William C. Randolph. 1992. "Household Income Mobility During the 1980s: A Statistical Assessment Based on Tax Return Data," *Tax Notes* Website (June 1).
Hugo, Victor. 1888. *Les Misérables*. Paris: Hetzel Publishers.
Hume, David. 1978 [1740]. *Treatise of Human Nature*. Oxford: Oxford University Press.
Kleinig, John. 1971. "The Concept of Desert," *American Philosophical Quarterly* 8: 71–8.
Kukathas, Chandran. 2003. "Responsibility for Past Injustice: How to Shift the Burden," *Politics, Philosophy, and Economics* 2: 165–90.
Kummer, Hans. 1991. "Evolutionary Transformation of Possessive Behavior," *Journal of Social Behavior and Personality* 6: 75–83.

Lacey, A. R. 2001. *Robert Nozick.* Princeton, NJ: Princeton University Press.
Lerman, Robert I. 1996. "The Impact of the Changing U.S. Family Structure on Child Poverty and Income Inequality," *Economica* 63: 119–39.
Lerner, Abba. 1970. *The Economics of Control.* New York: Augustus M. Kelley Publishers.
Levitan, Sar A. 1990. *Programs in Aid of the Poor.* Baltimore, MD: Johns Hopkins University Press.
Locke, John. 1960 [1690]. *Two Treatises of Government.* Cambridge: Cambridge University Press.
Lomasky, Loren. 2001. "Nozick on Utopias," in Schmidtz, ed. *Robert Nozick.* 59–82. New York: Cambridge University Press.
Lomasky, Loren. 2005. "Libertarianism at Twin Harvard," *Social Philosophy and Policy* 22: 178–99.
Louden, Robert B. 1992. *Morality and Moral Theory.* New York: Oxford University Press.
Mandle, Jon. 2000. *What's Left of Liberalism?* Lanham, MD: Lexington.
Maslow, Abraham. 1970. *Motivation and Personality.* New York: Harper & Row.
McCloskey, Deirdre. 1985. *The Rhetoric of Economics.* Madison: University of Wisconsin Press.
McConnell, Terrance. 1993. *Gratitude.* Philadelphia: Temple University Press.
McMurrer, Daniel P., Mark Condon, and Isabel V. Sawhill. 1997. *Intergenerational Mobility in the United States.* Washington, DC: Urban Institute.
McKerlie, Dennis. 1989. "Equality and Time," *Ethics* 99: 475–91.
McNeil, John. 1998. "Changes in Median Household Income: 1969 to 1996," U.S. Department of Commerce Website, *Current Population Reports, Special Studies P23–196.* Washington, DC: U.S. Government Printing Office.
Mill, John Stuart. 1974 [1859]. *On Liberty.* Harmondsworth: Penguin.
Mill, John Stuart. 1979 [1861]. *Utilitarianism.* Indianapolis, IN: Hackett.
Miller, David. 1976. *Social Justice.* Oxford: Oxford University Press.
Miller, David. 1999a. *Principles of Social Justice.* Cambridge, MA: Harvard University Press.
Miller, David. 1999b. "Justice and Global Inequality," in Hurrell and Woods, eds. *Inequality, Globalization, and World Politics.* 187–210. Oxford: Oxford University Press.
Miller, Fred D. 2001. "Sovereignty and Political Rights," in Otfried Höffe, ed. *Aristoteles Politik.* 107–19. Berlin: Akademie Verlag.
Miller, Richard W. 1992. "Justice as Social Freedom," in Beehler, Szabados, and Copp, eds. *On the Track of Reason: Essays in Honor of Kai Nielsen.* 37–55. Boulder, CO: Westview.
Miller, Richard W. 2002. "Too Much Inequality," *Social Philosophy and Policy* 19: 275–313.
Morris, Christopher. 1991. "Punishment and Loss of Moral Standing," *Canadian Journal of Philosophy* 21: 53–79.
Morris, Christopher. 1998. *An Essay On the Modern State.* Cambridge: Cambridge University Press.
Nagel, Thomas. 1989. "Rawls on Justice," in Daniels, eds. *Reading Rawls.* 1–16. Stanford, CA: Stanford University Press.

Nagel, Thomas. 1991. *Equality and Partiality.* Oxford: Oxford University Press.
Nagel, Thomas. 1997. "Justice and Nature," *Oxford Journal of Legal Studies* 2: 303–21.
Narveson, Jan. 1994. "Review of Temkin's *Inequality*," *Philosophy and Phenomenological Research* 56: 482–86.
Narveson, Jan. 1995. "Deserving Profits," in Cowan and Rizzo, eds. *Profits and Morality.* 48–97. Chicago: University of Chicago Press.
Narveson, Jan. 1997. "Egalitarianism: Baseless, Partial, and Counterproductive," *Ratio* 10: 280–95.
Nietzsche, Friedrich. 1969 [1887]. *On the Genealogy of Morals.* New York: Vintage Books.
Norris, Floyd. 1996. "So Maybe It Wasn't the Economy," *New York Times* December 1.
Nozick, Robert. 1974. *Anarchy, State, and Utopia.* New York: Basic Books.
Olsaretti, Serena. 2004. *Liberty, Desert, and the Market.* Cambridge: Cambridge University Press.
Piketty, Thomas and Emmanuel Saez. 2004. "Income Inequality in the United States, 1913–2002," an on-line update of "Income Inequality in the United States, 1913–1998," *Quarterly Journal of Economics* 118(2003): 1–39.
Price, Terry L. 1999. "Egalitarian Justice, Luck, and the Costs of Chosen Ends," *American Philosophical Quarterly* 36: 267–78.
Rachels, James. 1997. "What People Deserve," in *Can Ethics Provide Answers?* 175–97. Lanham, MD: Rowman and Littlefield.
Radzik, Linda. 2004. *Making Amends.* Texas A&M University, Unpublished manuscript.
Rakowski, Eric. 1991. *Equal Justice.* Oxford: Oxford University Press.
Rawls, John. 1971. *A Theory of Justice.* Cambridge, MA: Harvard University Press.
Rawls, John. 1996. *Political Liberalism.* New York: Columbia University Press.
Rawls, John. 1999a. *A Theory of Justice.* Revised ed. Cambridge, MA: Harvard University Press.
Rawls, John. 1999b. *Collected Papers.* ed. S. Freeman. Cambridge, MA: Harvard University Press.
Rawls, John. 1999c. *Law of Peoples.* Cambridge, MA: Harvard University Press.
Rawls, John. 2001. *Justice as Fairness: A Restatement.* Cambridge, MA: Harvard University Press.
Rector, Robert and Rea S. Hederman. 1999. "Income Inequality: How Census Data Misrepresent Income Distribution," *Report of the Heritage Center for Data Analysis.* Washington, DC: Heritage Foundation.
Roemer, John E. 2002. "Egalitarianism Against the Veil of Ignorance," *Journal of Philosophy* 99: 167–84.
Rose, Carol. 1985. "Possession As the Origin Of Property," *University of Chicago Law Review* 52: 73–88.
Rovane, Carol A. 1998. *The Bounds of Agency: An Essay in Revisionary Metaphysics.* Princeton, NJ: Princeton University Press.
Samuelson, Paul. 1973. *Economics,* 9th ed. New York: McGraw-Hill Publishers.
Sanders, John T. 2002. "Projects and Property," in Schmidtz, ed. *Robert Nozick.* 34–58. New York: Cambridge University Press.

Sayre-McCord, Geoffrey. 1996. "Hume and the Bauhaus Theory of Ethics," *Midwest Studies* 20: 280–98.
Scheffler, Samuel. 1992. "Responsibility, Reactive Attitudes, and Liberalism in Philosophy and Politics," *Philosophy and Public Affairs* 21: 299–323.
Schmidtz, David. 1990a. "Justifying the State," *Ethics* 101: 89–102.
Schmidtz, David. 1990b. "When Is Original Appropriation *Required?*" *Monist* 73: 504–18.
Schmidtz, David. 1992. "Rationality Within Reason," *Journal of Philosophy* 89: 445–66.
Schmidtz, David. 1994. "The Institution of Property," *Social Philosophy & Policy* 11: 42–62.
Schmidtz, David. 1995. *Rational Choice and Moral Agency*. Princeton, NJ: Princeton University Press.
Schmidtz, David and Elizabeth Willott, eds. 2003. "Reinventing the Commons: An African Case Study," *University of California at Davis Law Review* 36: 203–32.
Sen, Amartya. 1992. *Inequality Reexamined*. Cambridge, MA: Harvard University Press.
Sher, George. 1987. *Desert*. Princeton, NJ: Princeton University Press.
Sher, George. 1997. "Ancient Wrongs and Modern Rights," in *Approximate Justice*. 15–27. Lanham, MD: Rowman and Littlefield.
Shue, Henry. 2002. "Global Environment and International Inequality," in Schmidtz and Willott, eds. *Environmental Ethics: What Really Matters, What Really Works* 394–404. New York University Press.
Simmons, A. John. 1979. *Moral Principles and Political Obligations*. Princeton, NJ: Princeton University Press.
Smith, Adam. 1982 [1759]. *The Theory of Moral Sentiments*. Indianapolis, IN: Liberty Fund.
Spector, Horacio. 1992. *Autonomy and Rights*. Oxford: Oxford University Press.
Stark, Cynthia A. 2004. "How To Include the Severely Disabled in a Contractarian Theory of Justice," University of Utah, Unpublished manuscript.
Taylor, Charles. 1985. "Atomism," in *Philosophy and the Human Sciences: Philosophical Papers, vol.* 2. Cambridge: Cambridge University Press.
Taylor, Charles. 1995. *Philosophical Arguments*. Cambridge, MA: Harvard University Press.
Temkin, Larry S. 1993. *Inequality*. New York: Oxford University Press.
Thomson, Judith. 1976. "Killing, Letting Die, and the Trolley Problem," *Monist* 59: 204–17.
Tomasi, John. 2001. *Liberalism Beyond Justice*. Princeton, NJ: Princeton University Press.
U.S. Census Bureau. 2005. "Poverty Thresholds: 2004," *January Current Population Survey*. Washington, DC: U.S. Government Printing Office.
U.S. Census Bureau. 2003. *Current Population Reports, P60-221, Income In the United States: 2002*. Washington, DC: U.S. Government Printing Office.
U.S. Department of the Treasury, Office of Tax Analysis. 1992. "Household Income Changes Over Time: Some Basic Questions and Facts," *Tax Notes, August 24, 1992*: Washington, D.C: U.S. Government Printing Office.

Waldron, Jeremy. 1989. "The Rule of Law in Contemporary Liberal Theory," *Ratio Juris* 2: 79–96.
Waldron, Jeremy. 1992. "Superseding Historic Injustice," *Ethics* 103: 4–28.
Waldron, Jeremy. 1995. "The Wisdom of the Multitude: Some Reflections on Book 3, Chap. 11 of Aristotle's *Politics*," *Political Theory* 23: 563–84.
Walzer, Michael. 1983. *Spheres of Justice.* New York: Basic Books.
Wellman, Christopher Heath. 1999. "Gratitude as a Virtue," *Pacific Philosophical Quarterly* 80: 284–300.
Wellman, Christopher Heath. 2002. "Justice," in Simon, ed. *The Blackwell Guide to Social and Political Philosophy.* 60–84. Malden: Blackwell.
Williams, Bernard. 1985. *Ethics and the Limits of Philosophy.* Cambridge, MA: Harvard University Press.
Willott, Elizabeth. 2002. "Recent Population Trends," in Schmidtz and Willott, eds. *Environmental Ethics: What Really Matters, What Really Works.* 274–83. New York: Oxford University Press.
Wittgenstein, Ludwig. 1958. *Philosophical Investigations*, 3rd ed. Anscombe, trans. New York: MacMillan.
Young, Iris Marion. 1990. *Justice and the Politics of Difference.* Princeton, NJ: Princeton University Press.
Zaitchik, Alan. 1977. "On Deserving to Deserve," *Philosophy and Public Affairs* 6: 370–88.

Index